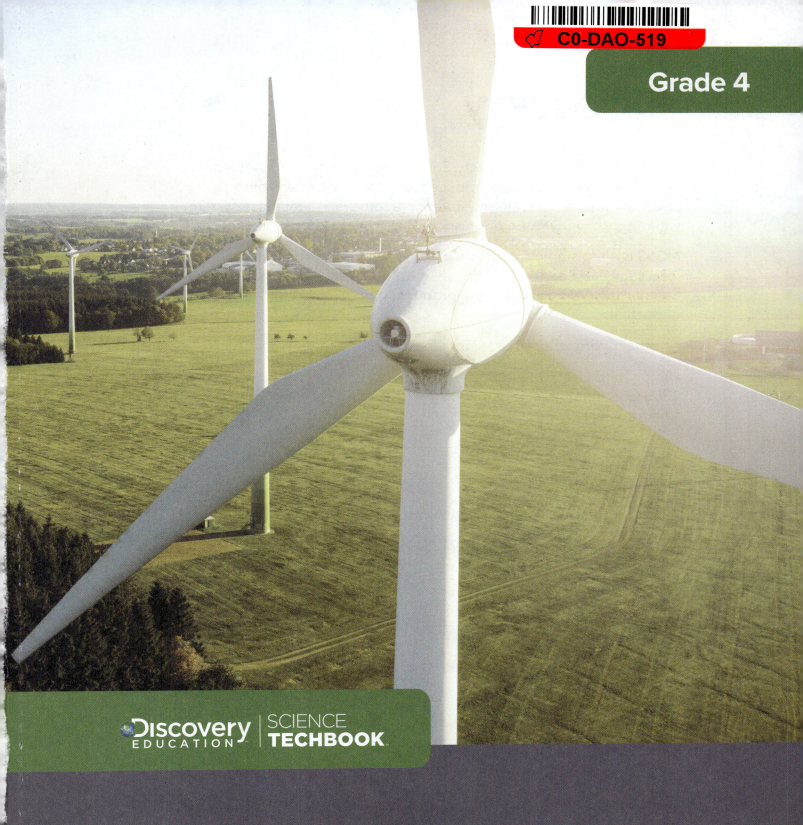

Grade 4

DISCOVERY EDUCATION | SCIENCE TECHBOOK

California
Unit 2
Energy Resources

To obtain permission(s) or for inquiries, submit a request to:

Discovery Education, Inc.
4350 Congress Street, Suite 700
Charlotte, NC 28209
800-323-9084
Education_Info@DiscoveryEd.com

ISBN 13: 978-1-68220-546-4

Printed in the United States of America.

1 2 3 4 5 6 7 8 9 10 CJH 23 22 21 20 19   A

Acknowledgments

Acknowledgement is given to photographers, artists, and agents for permission to feature their copyrighted material.

Cover and inside cover art: r.classen / Shutterstock.com

Table of Contents

Concept 2.3 Renewable Energy Resources

Concept 2.4 Energy and the Environment

Unit Wrap-Up

Grade 4 Resources

Dear Parent/Guardian,

This year, your student will be using Science Techbook™, a comprehensive science program developed by the educators and designers at Discovery Education and written to the California Next Generation Science Standards (NGSS). The California NGSS expect students to act and think like scientists and engineers, to ask questions about the world around them, and to solve real-world problems through the application of critical thinking across the domains of science (Life Science, Earth and Space Science, Physical Science).

Science Techbook is an innovative program that helps your student master key scientific concepts. Students engage with interactive science materials to analyze and interpret data, think critically, solve problems, and make connections across science disciplines. Science Techbook includes dynamic content, videos, digital tools, Hands-On Activities and labs, and game-like activities that inspire and motivate scientific learning and curiosity.

You and your child can access the resource by signing in to www.discoveryeducation.com. You can view your child's progress in the course by selecting the Assignment button.

© Discovery Education | www.discoveryeducation.com

Science Techbook is divided into units, and each unit is divided into concepts. Each concept has three sections: Wonder, Learn, and Share.

Units and Concepts Students begin to consider the connections across fields of science to understand, analyze, and describe real-world phenomena.

Wonder Students activate their prior knowledge of a concept's essential ideas and begin making connections to a real-world phenomenon and the **Can You Explain?** question.

Learn Students dive deeper into how real-world science phenomenon works through critical reading of the Core Interactive Text. Students also build their learning through Hands-On Activities and interactives focused on the learning goals.

Share Students share their learning with their teacher and classmates using evidence they have gathered and analyzed during Learn. Students connect their learning with STEM careers and problem-solving skills.

Within this Student Edition, you'll find QR codes and quick codes that take you and your student to a corresponding section of Science Techbook online. To use the QR codes, you'll need to download a free QR reader. Readers are available for phones, tablets, laptops, desktops, and other devices. Most use the device's camera, but there are some that scan documents that are on your screen.

For resources in California Science Techbook, you'll need to sign in with your student's username and password the first time you access a QR code. After that, you won't need to sign in again, unless you log out or remain inactive for too long.

We encourage you to support your student in using the print and online interactive materials in Science Techbook, on any device. Together, may you and your student enjoy a fantastic year of science!

Sincerely,

The Discovery Education Science Team

Unit 2
Energy Resources

Raging Water

The sight of millions of tons of water flowing down a river and over a waterfall is impressive. All that water has a lot of kinetic energy. By the end of this unit, you will be able to describe how that energy can be turned into useful electricity. You will also be able to evaluate how obtaining that energy impacts the environment.

Quick Code:
ca4456s

Raging Water

Discovery EDUCATION

Think About It

Look at the photograph. **Think** about the following questions.

- How do we get electricity and fuel to run cars and power electronic devices?

- How does human use of natural resources affect the environment?

Waterfall

Solve Problems Like a Scientist

Quick Code:
ca4457s

Unit Project: Dam Impacts

In this project, you will use what you know about energy and the environment to assess the positive and negative impacts of building a dam on the surrounding environment.

Hoover Dam

SEP Planning and Carrying Out Investigations

CCC Stability and Change

Ask Questions About the Problem

You are going to evaluate the positive and negative impacts that building the Hoover Dam had on the surrounding environment, including humans, wildlife, and the landscape. Then, you will research solutions to one of the negative impacts you identify. **Write** some questions you can ask to learn more about the problem. As you learn about how the ways people use energy affect the environment, **write** answers to your questions.

Devices and Energy

Student Objectives

By the end of this lesson:

☐ I can develop models based on observations that describe how everyday devices transform, yet conserve, energy.

☐ I can use observations and evidence to explain how energy is transferred from place to place by sound, light, heat, and electricity.

Key Vocabulary

☐ chemical energy   ☐ generate

☐ conduction   ☐ photosynthesis

☐ conserve   ☐ radiant energy

☐ Earth   ☐ remote

☐ energy source   ☐ sound

☐ energy transfer   ☐ sun

Quick Code:
ca4458s

Activity 1
Can You Explain?

How does energy from the sun power an electric toaster?

_____

_____

_____

_____

_____

_____

_____

_____

_____

_____

_____

Quick Code:
ca4460s

Activity 2

Ask Questions Like a Scientist

Quick Code:
ca4461s

Energy in Remote-Controlled Cars

Read the text and **look** at the image. Then, **complete** the activity that follows.

Energy in Remote-Controlled Cars

Many toys can be operated remotely. Remote-controlled cars, trucks, planes, and boats are fun to use. Drones are becoming popular playthings, as are complex remote-controlled robots. All these devices need energy to make them move and do tasks such as turning corners, moving remote arms, or operating cameras.

Let's Investigate Energy in Remote-Controlled Cars

Where do they get this energy from? All of these devices use electricity. Batteries are their onboard **energy source**. When the batteries are exhausted, they must be recharged or replaced with new ones. That's easy. Simply plug the device into the nearest charger or run down to the hardware store. But sometimes that isn't possible.

Write some questions you would like to investigate about the devices you use in your daily lives and the energy that they use.

How is the satillite built for internet connection?

How many sources are there other than satillites?

How does an outlet work?

SEP Asking Questions and Defining Problems

CCC Energy and Matter

Analyze Like a Scientist

Mars Rover

Quick Code:
ca4462s

Look at the picture and **read** the text. Then, **complete** the activity that follows.

Mars Rover

Mars never gets closer to **Earth** than about 54 million kilometers. That's a long way. It takes a spacecraft about six months, usually longer, to get there. Over the past few decades, humans have sent many missions to Mars. None of these missions included people; they all used different types of remotely operated vehicles or robots. These robots have performed a variety of jobs. The most famous of these is the Mars Rover, which travels on the surface of the planet. Scientists have remotely operated four rovers on the surface of Mars.

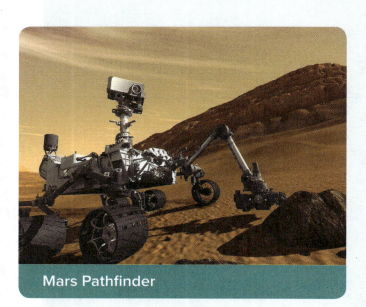

Mars Pathfinder

Like remote-controlled toys, these rovers need energy. They also use electricity. However, the rovers are too far from a local store or socket on Earth to use the same types of batteries as those found in toys. They can't just plug into the nearest Mars rock! What energy sources could they use?

List possible ways the Mars Rover gets its energy.

A place I think is since Mars is one big gas rock I think it's from gas. Maybe the Mars Rover runs on gas.

Activity 4
Observe Like a Scientist

How to Build a Rover

Watch the video and **look** for different systems or parts on the Mars Rover that have specific functions. Then, **discuss** how the rover investigates its surroundings.

Quick Code:
ca4463s

How to Build a Rover

 **Talk Together**

What specialized instruments does the Rover need? What power source does it need?

CCC   Structure and Function

Activity 5

Evaluate Like a Scientist

What Do You Already Know About Devices and Energy?

Generating Electricity

Look at the situations. **Highlight** those that <mark>generate</mark> electricity.

- sun on solar panels
- candles
- magnifying glass
- wind turbine
- hydroelectric dam
- ~~electrical wall socket~~
- balloon rubbed on a head
- ~~unconnected wires~~

Using Electricity

Draw a line to **match** each set of devices to the form of energy it shows.

radio, car horn, doorbell	Motion
fan, motorized toy, food mixer	Heat
flashlight, desk lamp, LED	Sound
toaster, oven, curling iron	Light

What Do Devices Do with Energy?

Activity 6

Observe Like a Scientist

Quick Code:
ca4465s

Energy Input and Output

Look at the pictures. Then, **discuss** the questions.

 Talk Together

What is the source of energy, or energy input, for each device?
What is the energy output?

© Discovery Education | www.discoveryeducation.com ● Images: (a) Romwel Dotig / EyeEm / Getty Images, (b) Gary Ombler / Dorling Kindersley / Getty Images, (c) Science Photo Library / Science Photo Library / Getty Images, (d) Emilija Manevska / Moment / Getty Images (e) Icon made by Freepik from www.flaticon.com

CCC   **Energy and Matter**

Activity 7

Think Like a Scientist

Quick Code:
ca4466s

Energy and Everyday Devices

In this investigation, you will determine the energy input and output of several devices.

What materials do you need?

- A variety of 10 common devices

- Basic tools such as screwdrivers (optional)

- Energy of Everyday Devices sheet

What Will You Do?

1. Analyze each device.

2. Determine the energy input for the device.

3. Determine the energy output for the device.

4. Record your observations on the Energy of Everyday Devices sheet.

| **SEP** | **Constructing Explanations and Designing Solutions** |
| **CCC** | **Energy and Matter** |

Think About the Activity

How did you determine the type of energy that went into the use of
each device?

_____

_____

_____

_____

How did you determine the type of energy that came out of each device
as it was used?

_____

_____

_____

_____

Does all of the energy that goes into each device come out as part of
its function, or is some of the energy wasted? Support your answer with
examples.

_____

_____

_____

_____

© Discovery Education | www.discoveryeducation.com • Images: (a) Romwel Dotig / EyeEm / Getty Images, (b) Krista Kennell / Shutterstock.com

Activity 8

Observe Like a Scientist

Quick Code:
ca4467s

The Conservation of Energy

Watch the video several times. After you have watched, **answer** the questions that follow.

The Conservation of Energy

What is the definition of the phrase *conservation of energy*?

_____

_____

What are the different types of energy that are converted when a light bulb is turned on?

_____

_____

CCC **Energy and Matter**

Quick Code:
ca4468s

Activity 9
Analyze Like a Scientist

Energy Conservation

Read the text and **look** at the diagram. Then, **complete** the activity that follows.

Energy Conservation

Energy is **conserved**. It is neither created nor destroyed. All the energy that goes into a device must eventually leave it, in the same or a different form. So, these devices have energy that goes in and energy that goes out. We call these energy inputs and energy outputs.

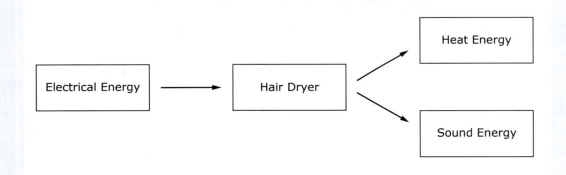

Take the hair dryer as an example. The energy input to the hair dryer comes through the cord as electrical energy. Inside the dryer, that energy is partly changed into other forms. Thermal energy, **sound**, and kinetic energy (from the fan and the moving air) leave the dryer. These are its energy outputs.

CCC **Energy and Matter**

© Discovery Education | www.discoveryeducation.com • Images: Remitski Ivan / Shutterstock.com

Sometimes, energy enters a device and is stored inside it for a while. A cell phone is one example. Energy enters the device as electrical energy. It is stored in the battery of the phone as **chemical energy**. When a phone is on or in use, the phone changes some of this stored energy. Can you think of how the phone uses this stored energy?

List the different ways a cell phone uses its stored energy.

_____

_____

_____

_____

_____

_____

_____

_____

Activity 10
Evaluate Like a Scientist

Quick Code:
ca4469s

Electric Transformations

Write the correct terms from the word bank in the empty boxes to show the energy type used at that part of the system. Not all terms will be used.

electrical energy   sound energy

light energy   motion energy

[   ]

Battery

↓

[   ]

↓

[Mobile Phone]

↙   ↘

[   ]   [   ]

SEP **Constructing Explanations and Designing Solutions**

Activity 11

Analyze Like a Scientist

Moving Energy

Read the text. Then, **complete** the activity that follows.

Moving Energy

Think of how you turn food energy into energy for running or reading. Where did the food's energy come from? Nuclear energy in the **sun** is converted to **radiant energy** such as light. Some of this radiant energy strikes Earth. Plants turn light energy into chemical energy as they make sugars. This process is called **photosynthesis**. Animals eat the plants, and the energy moves up the food chain. Eventually, some of that energy might land on your plate as eggs or peas!

Energy moves and transforms in many other ways, too. In fact, for energy to do anything, it must move from place to place. The electrical energy that powers a hair dryer reaches it along an electric cord that is made of copper. The electrical energy comes from a power plant of some type. Perhaps it burned coal or gas to make this electrical energy. But where did this energy originally come from?

CCC Energy and Matter

Choose two sentence starters from the phrase bank and use them to write sentences about what you have read.

I don't get this part...	This is really saying...	This reminds me of...
I wonder if...	This is confusing because...	The basic idea is...
What do you think...	This makes sense now...	An example is...
I predict that...	I think that...	My understanding is...

This is confusing because how does the sugar get into the plant. What does chemical energy do? My understanding is that the sun makes the sugar, puts it in the plant, the animals eats the plant, people kill the animal, put it in the supermarket, our parents buy it, feeds it to us, and get sun tn energy in our bodies?

Activity 12

Observe Like a Scientist

Food Energy

Watch the video and **look** for producers and consumers in a food chain. Then, **talk** together about energy chains.

Video

Food Energy

Talk Together

What energy chains have at least three steps?

Where Does the Energy We Use Come From and Go To?

Activity 13
Analyze Like a Scientist

Energy Chains

Read the text and **look** at the diagrams. Then, **share** what you've read with your partner.

Energy Chains

Most energy we use is made inside the sun. This was what we found in the food chain. A food chain is one type of energy chain. Just as it maps the transfer of energy from the sun through plants to animals, we can draw energy chains that track energy from the sun to different devices.

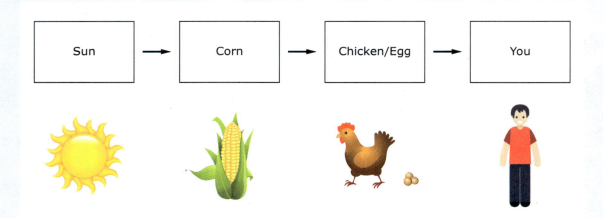

| Sun | → | Corn | → | Chicken/Egg | → | You |

SEP Developing and Using Models

Let's start with a simple example: heating a pan of water on a camp fire. The energy to heat the water comes from the burning wood. Where does the chemical energy in the wood come from? Well, the wood was originally part of a tree, and trees make their food using energy from the sun. Put this all together and we have a chain from the sun to the thermal energy of the hot water.

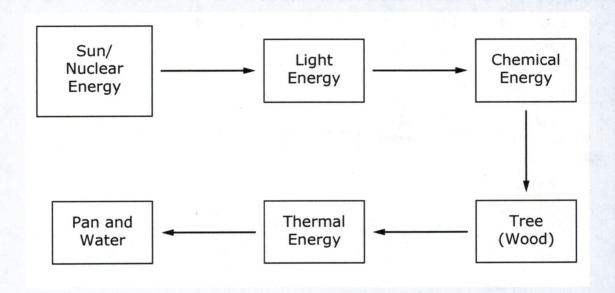

Energy Chains *cont'd*

The energy chain for a hair dryer is more difficult. We already have many of the links. We have traced the energy backward to the power plant. If that power plant burned coal, a form of chemical energy, then we need just one more link. One missing link is energy from the sun. Coal was formed millions of years ago from dead trees. Where did the trees get their energy from? You guessed it: sunlight. So now we can add to our earlier diagram about energy in a hair dryer.

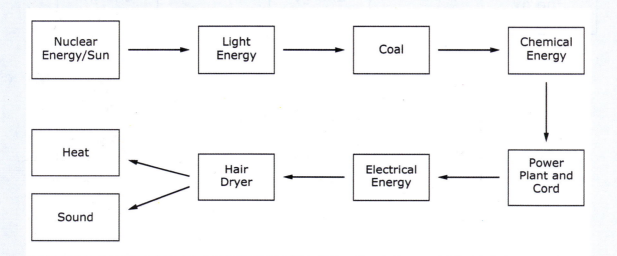

If we have a device and power it, not all energy that enters an energy chain reaches the device and gets used as we intend. At each link in the chain, some energy escapes as other forms. We can think of it as wasted energy. It still exists, but it gets transformed into energy we cannot use. Most of this energy gets turned into heat. It never disappears.

Quick Code:
ca4473s

Activity 14

Think Like a Scientist

Build an Energy Chain

In this investigation, you will model **energy transfer** pathways by creating an energy chain.

What Will You Do?

Use magazine pictures to **create** an energy chain. **Include** at least six pictures. **Label** each picture, the form of energy, and whether the energy is being transferred or transformed.

SEP Developing and Using Models　　**CCC** Energy and Matter

Think About the Activity

1. How can these types of models be used to track energy pathways?

   _____

   _____

   _____

   _____

   _____

   _____

2. How can these types of models be used to track the movement of energy through the environment?

   _____

   _____

   _____

   _____

   _____

   _____

3. What are the limitations of these types of models?

   _____

   _____

   _____

   _____

   _____

   _____

Activity 15

Evaluate Like a Scientist

Quick Code:
ca4474s

Complete the Energy Chain

Complete the energy chain. **Write** each word from the word bank in the correct box.

solar radiation   power plant   friction

chemical energy   heat

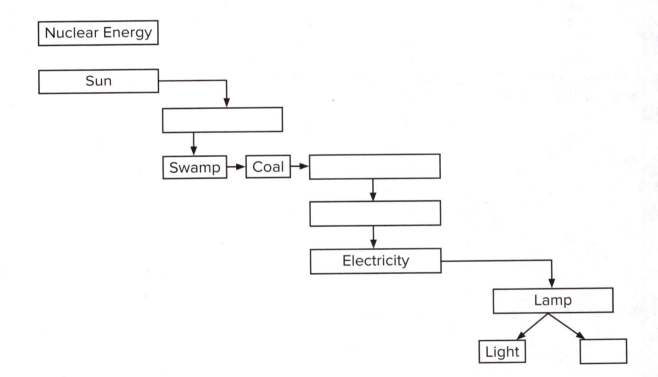

Nuclear Energy

Sun

Swamp → Coal →

Electricity

Lamp

Light

CCC **Energy and Matter**

Activity 16

Think Like a Scientist

Present Your Energy Chain

Quick Code:
ca4475s

In this investigation, you will share your energy chain.

Present your energy chain to your classmates. Then, **review** their energy chains. Finally, **answer** the questions that follow.

How can these types of models be used to track energy pathways?

_____

_____

_____

How can these types of models be used to track the movement of energy through the environment?

_____

_____

_____

_____

What are the limitations of these types of models?

_____

_____

_____

SEP   Developing and Using Models

© Discovery Education | www.discoveryeducation.com   ●   Image: Romwel Dotig / EyeEm / Getty Images

Activity 17

Record Evidence Like a Scientist

Quick Code:
ca4476s

Energy in Remote-Controlled Cars

Now that you have learned about energy, look again at the image of a remote-controlled car. You first saw this in Wonder.

Let's Investigate Energy in Remote-Controlled Cars

Talk Together

How can you describe the energy in a remote-controlled car now? How is your explanation different from before?

SEP   **Constructing Explanations and Designing Solutions**

© Discovery Education | www.discoveryeducation.com ● Images: (a) Romwel Dotig / EyeEm / Getty Images, (b) Hero Images / Hero Images / Getty Images. (c) Icon made by Freepik from www.flaticon.com

Look at the Can You Explain? question. You first read this question at the beginning of the lesson.

Can You Explain?

How does energy from the sun power an electric toaster?

Now, you will use your new ideas about energy to answer a question.

1. **Choose** a question. You can use the Can You Explain? question or one of your own. You can also use one of the questions that you wrote at the beginning of the lesson.

My Question

2. Then, use the graphic organizers on the next pages to help you answer the question.

To plan your scientific explanation, first **write** your claim. Your claim is a one-sentence answer to the question you investigated. It answers: What can you conclude? It should not start with yes or no.

My claim:

_____

_____

_____

_____

_____

_____

_____

Record the reasons and evidence to support your claim in the graphic organizer.

Reasoning

Evidence

Now, **write** your scientific explanation.

Energy from the sun powers an electric toaster by . . .

_____

_____

_____

_____

_____

_____

_____

_____

_____

_____

_____

_____

_____

_____

_____

_____

_____

_____

_____

Discovery
EDUCATION

STEM in Action

Analyze Like a Scientist

Careers and Energy in Systems

Quick Code:
ca4477s

Read the text and **complete** the activities that follow.

Careers and Energy in Systems

Many types of engineers and scientists have careers that require knowledge of energy in systems. Ecologists study energy in ecosystems. They examine how energy flows through food webs. Changes in the flow of energy can affect living things. Some ecologists study the movement of energy in extreme ecosystems.

Engineers use their understanding of energy in systems to design technologies that we use every day. Think about a cell phone or a computer. How does the screen get the energy it needs to produce a lighted display? How are the sounds produced? How does a phone carry the sound of your voice? Engineers must understand how to design parts of a system to change energy from one form to another.

CCC **Energy and Matter**

CCC **Systems and System Models**

Can you think of other careers that require an understanding of energy in systems?

_____

_____

_____

_____

_____

_____

Cell Phone Energy

Write the energy inputs and outputs in a cell phone in the correct column.

electrical energy   chemical energy   light

sound   heat

Inputs	Outputs	Both

Energy Problems

Think about the process that electronic engineers use when they develop electronics such as cell phones. What is one energy-related problem that they might encounter in the cell phone system? What steps of the engineering process do they likely use to solve the problem? **Use reasoning** to support your claim.

_____

_____

_____

_____

_____

_____

_____

_____

_____

_____

_____

_____

Activity 19

Evaluate Like a Scientist

Review:
Devices and Energy

Think about what you have read and seen in this lesson. **Write** down some core ideas you have learned. **Review** your notes with a partner. Your teacher may also have you take a practice test.

_____

_____

_____

_____

_____

_____

 ## Talk Together

Think about what you saw in Get Started. Use your new ideas to discuss energy, how it moves, and how it powers devices and humans.

SEP **Obtaining, Evaluating, and Communicating Information**

About Fuels

Student Objectives

By the end of this lesson:

- [ ] I can describe patterns in how different types of fossil fuels are formed and predict the properties and uses of different types of fossil fuels.

- [ ] I can show how energy flows from a fuel source to a transportation vehicle and is then emitted in the air.

- [ ] I can describe how the use of energy and fuels affects the environment.

Key Vocabulary

- [ ] fossil fuel
- [ ] fuel
- [ ] nonrenewable
- [ ] renewable

Quick Code:
ca4479s

Activity 1

Can You Explain?

How are fossil fuels formed and used?

_____

_____

_____

_____

_____

_____

_____

_____

_____

_____

_____

Quick Code:
ca4481s

Activity 2

Ask Questions Like a Scientist

Quick Code:
ca4482s

Fuels and Road Trips

Read the text and **look** at the photographs. Then, **complete** the activity that follows.

Fuels and Road Trips

They were an hour into the long road trip to Aunt Camila's. Unlike her younger brother Mateo, who was fast asleep, Juanita was getting bored. She looked over her mom's shoulder to see how fast they were going.

Let's Investigate Fuels and Road Trips

It was then that Juanita noticed how low the gas was. "Hey Mom, we're running out of gas, and there are no gas stations on this interstate."

Her mom glanced down at the gauge. "Wow. I'll check out the next exit; perhaps there's a gas station."

Mom Pumping Gas

At last, seven kilometers farther along the highway, a turnoff came into view. The car raced up the ramp and sputtered its way over to the nearest gas station. The sputtering woke Mateo up. He looked worried. "Are we there yet? What's wrong with the car?"

"The car is out of gas."

The sputtering stopped, and the car slowly rolled across the gas station forecourt. "Are we going to make it?" Mateo gasped.

The car rolled to a stop close to the nearest gas pump. "We just made it," his mom said. "Another minute and we would have been pushing the car along the hard shoulder."

While their mom got out to fill the car, Mateo rolled down the window. His mom was fiddling with the pump and pressed one of the numbers on the front. Mateo loved to ask questions. "Hey Mom, why do we need gas? Why don't they make cars that don't need it? Then we wouldn't need to stop. We'd be at Aunt Camila's already."

Fuels and Road Trips *cont'd*

By now, his mom was pumping the gas into the **fuel** tank. "Cars need energy to run. The car burns the gas in its engine, and the engine turns the wheels. No gas, no movement."

"But why?" said Mateo. "Can't we run the car on something else? Could we get a car that runs on sunshine?"

His mom laughed. "Well, I don't think they sell them yet. Anyhow, how would we drive at night?"

Think about: What are fuels, and what are they used for? Was Mateo's mom right in the way she described what gas does inside a car? Do you think Mateo's idea about a car that runs on sunshine is a good one? In this concept, you will learn about fuels and some other sources of energy that we use.

Write other questions you would like to investigate about different types of fuels, where they come from, and how we use them.

_____

_____

_____

_____

SEP Asking Questions and Defining Problems

CCC Energy and Matter

Activity 3

Evaluate Like a Scientist

Quick Code:
ca4483s

What Do You Already Know About Fuels?

Nobody's Fuel

Write whether each form of energy in the word bank is fuel energy or another type of energy.

gasoline	solar	wind
animal dung	charcoal	wood
olive oil	ethanol	geothermal
coal	natural gas	uranium

Fuel Energy	Other Energy

Which Statement?

Circle the statement that describes the relationship between natural resources and energy.

Natural resources and energy are both examples of matter.

We can use up all the energy in natural resources.

Energy can be produced from natural resources.

All natural resources can easily be replaced.

What Are the Different Types of Fuels?

Activity 4

Analyze Like a Scientist

Types of Fuel

Quick Code:
ca4484s

Read the text. As you read, **record** details about biofuels and fossil fuels in the graphic organizer at the end.

Types of Fuel

Fuels are substances that, when burned, release thermal energy. Wood is the most ancient fuel and is still widely used throughout the world. A wide variety of plant and other materials are used for fuels. Because they are made from living things, they are called biofuels. For example, charcoal, made from wood, is an important fuel.

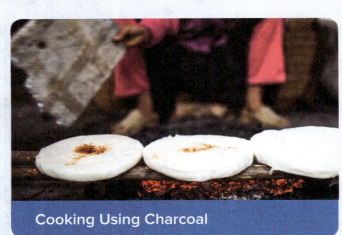

Cooking Using Charcoal

SEP Obtaining, Evaluating, and Communicating Information

CCC Energy and Matter

In parts of Asia, animal dung is often used for cooking and heating.

Some plants can be turned into liquid fuels. For example, switch grass, wood chips, and corn all can be used to make a liquid fuel called ethanol.

Corn Flowers

Ethanol can be used much like gasoline. If we trace back to where the energy in these fuels comes from, we find that they started with light energy from the sun. All of these fuels are continually being burned but are also being renewed as plants grow. If managed carefully, these fuels could last forever. For this reason, they are called **renewable** fuels.

Fossil fuels are fuels that were formed from the remains of plants and animals that lived millions of years ago. Over a very long time, these remains built up and became buried under Earth's surface. For example, around 300 million years ago, large sections of Earth's continents were covered in swamps. When the trees and plants in these swamps died, they got covered in mud and sand. Eventually the decayed or broken-down plants were covered in hundreds of meters of mud and rock. Earth's heat and pressure turned these remains into coal. Fuels like coal are formed mainly from ancient plants, while fuels like oil and gas form mostly from tiny, ancient sea animals. Coal, gasoline, and natural gas are all examples of fossil fuels.

Fossil fuel, such as coal, oil, and gas, form very slowly over millions of years. This means that we use them up much faster than they are formed. Since it takes millions of years for plant and animal remains to become fossil fuels, they can be used up far more quickly than they took to produce.

Types of Fuel *cont'd*

Coal Train

For practical purposes, once we use fossil fuels, they are gone. They cannot be easily renewed. For this reason, fossil fuels are said to be **nonrenewable**.

Coal is dug out of the ground. This process is called mining. Sometimes tunnels are used to get at the coal. In other places, where the coal is nearer the surface, big pits are dug. Oil is a liquid. It is found in some rocks. We get the oil by drilling a pipe down to it.

The oils squirts or is pumped up through this pipe. Drilling is also used to get natural gas.

The oil that comes out of the ground is usually a thick and gooey mixture. It is called crude oil. For it to be useful as fuel, crude oil must be processed in a factory. Factories that process crude oil are called refineries. In refineries, the crude oil is separated into useful fuels. These fuels include liquefied petroleum gas (LPG), gasoline, kerosene, and paraffin wax. The process also makes other substances, such as lubricating oil and bitumen.

Oil Rig

| **Discovery** EDUCATION

	Biofuels	Fossil Fuels
Definition	Based on what I read Biofuels are made from living things. Bio means life thats also how I know what it means.	In the text it says Fossil fuels are fuels that are made from remains of animals. and animals
Examples	On page 56 it says that charcoal is an important fuel.	The author wrote the coal, gasaline, and natural gas are examples of fossil fuels.
Renewable or Nonrenewable	Biofuels are renewable because there is still living things around us so bio fuels are renewable	Fossil Fuels are nonrenewable once the are gone they are gone.

Activity 5
Analyze Like a Scientist

A Trip Back in Time

Read the text. Then, **complete** the graphic organizer that follows.

A Trip Back in Time

Let's take a trip back in time. Imagine that you visit Earth 350 million years ago. You would see that Earth looked very different from today. There was only one large continent on Earth called Pangaea.

Swamps covered the continent of Pangaea. Huge trees and plants with large leaves grew in these swamps. As the trees and other plants died, they slowly sank to the bottom of the swamps. Here, the dead plants decayed slowly over time. Sand and clay piled up on top of the decaying plants.

The remains of ancient swamps are a major source of coal.

SEP **Developing and Using Models**

The sand and clay formed a layer over the decaying plants. This layer slowly turned into hard rock. Over time, other layers of sand and clay formed. Each layer turned into rock.

The layers of rock pressed down and squeezed all the water out of the decaying plants. Over millions of years, the plants turned into coal.

Coal is an important fossil fuel on Earth.

Coal is a fossil fuel. A fossil fuel forms in the earth from the remains of dead plants and animals. It is a source of energy. When people burn coal, it releases energy. They can use this energy to heat their homes, cook their food, or make electricity. They can also use the energy from coal to power some trains, ships, and other vehicles.

Coal is still forming on Earth today. However, you will have to visit Earth 350 million years from now before these fossil fuels are ready to use.

Develop a flowchart to outline the process by which coal was formed.

Activity 6

Observe Like a Scientist

Plants as Fuels

Watch the video and **look for** different types of fuels and how we use them.

Quick Code:
ca4486s

Plants as Fuels

Talk Together

Now, talk together about the four fuels described in the video (crude oil, coal, peat, and wood) and how they are used.

Activity 7
Observe Like a Scientist

Fossil Fuels

Quick Code:
ca4487s

Complete the interactive. Then, **complete** the graphic organizer that follows.

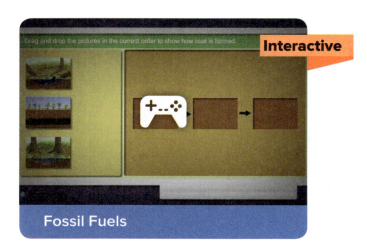

Interactive

Fossil Fuels

Stages of Coal Formation		
Stage 1	**Stage 2**	**Stage 3**

CCC **Energy and Matter**

© Discovery Education | www.discoveryeducation.com ● Image: Mathias Genterczewsky / EyeEm / Getty Images

Analyze Like a Scientist

Two Things That Do Not Mix

Read the text. Then, **answer** the questions that follow.

Two Things That Do Not Mix

Oil and water are very different. In fact, they are so different that they will separate if you try to mix them. Oil and water are also different in another way.

Oil comes from deep in the ground. Scientists think that oil formed from sea creatures called diatoms. Diatoms are tiny creatures that are no larger than the head of a pin. As these diatoms died, their remains settled on the ocean floor. They became covered with layers of sediment and rock.

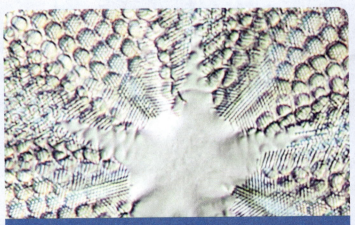

The shells of tiny diatoms are beautiful under a microscope.

Two Things That Do Not Mix *cont'd*

Over many millions of years, the sediment and rock built up more and more layers. All of these layers pressed down on the buried diatoms. By pressing down, the layers created great pressure and heat. This pressure and heat slowly turned the remains into oil.

Oil is a **nonrenewable resource**.

A nonrenewable resource is a natural material that is used faster than it can be replaced. We use oil faster than new oil can form. Therefore, we must use oil very wisely because we could run out of it.

Oil is nonrenewable. To get more, we must use oil rigs like this one.

Discovery EDUCATION

Unlike oil, water is a renewable resource. A renewable resource is a natural material that can be replaced soon after it is used. Even though water is renewable, we still must be careful when we use it. We should not waste or pollute water. If we do, then water may not be replaced as quickly as we need it.

Oil and water are very different. However, we must use them both very wisely.

Moving water can be used to produce energy.

What are some ways we could conserve these resources?

_____

_____

_____

_____

Why is water considered a renewable resource?

_____

_____

_____

Activity 9

Evaluate Like a Scientist

Quick Code:
ca4489s

Fossil Fuel Formation

Write the steps involved in the formation of fossil fuels in the correct order.

Remains changed to become coal, oil, and natural gas.

Remains were buried.

Living things that lived a long time ago died.

Heat and pressure affected the remains.

_____

_____

_____

_____

_____

_____

_____

CCC **Cause and Effect**

CCC **Energy and Matter**

What Are Fossil Fuels Used For?

Activity 10

Think Like a Scientist

Living without Electricity

In this activity, you will document your experience of spending some time without using electricity.

Quick Code:
ca4490s

What materials do you need?

- Notebook

- Pen or pencil

What Will You Do?

Choose a minimum of two hours to NOT use electricity. **Write** about your experience.

_____

_____

_____

_____

_____

_____

© Discovery Education | www.discoveryeducation.com ● Image: Mathias Genterczewsky / EyeEm / Getty Images

Think About the Activity

How long were you able to go without using electricity?

_____

_____

_____

What types of devices would you normally have used during this period of time? What did you do instead?

_____

_____

_____

_____

How did you feel during and after this experience? Do you feel that you normally take electricity for granted?

_____

_____

_____

_____

What can you do at home to conserve fuels and avoid wasting electricity?

_____

_____

_____

_____

Activity 11

Analyze Like a Scientist

Quick Code:
ca4491s

Using Fossil Fuels

Read the text. Then, **complete** the activity on the following page.

Using Fossil Fuels

You already know that gasoline is used to provide energy to make cars move. But what about the other fuels made from oil? And how are coal and natural gas used?

In the United States, most coal is used to generate electricity. Electricity is generated in a power plant. The coal is burned to release thermal energy. This is used to heat water to make steam. The steam is used to turn a device called a turbine. This kinetic energy is used to turn a generator. A generator transforms kinetic energy into electrical energy. The electrical energy travels down wires to homes and businesses. The chances are that, if you flip a light switch, the electricity you are using to make the bulb light comes, in part, from burning coal.

Substances made from crude oil are used in different ways. Gasoline is used in vehicles. The main use for kerosene is as fuel for jet airplanes. Diesel, sometimes called fuel oil, is used mainly to power trucks, buses, trains, and ships.

SEP Developing and Using Models

CCC Energy and Matter

Some fuel oil is used to generate electricity. Some people use it to heat their homes in winter. Liquefied Petroleum Gas (LPG) also can be used to power vehicles. Some buses use LPG because it causes less pollution than diesel.

Natural gas is used for heating, cooking, and electricity generation. It is also used to fuel some vehicles. It is used to make plastics. In the last twenty years, natural gas has become a very important fuel in the United States. It is quickly replacing coal as the preferred fuel for generating electricity.

LPG

Draw a model showing the flow of energy from coal, through a power plant, and to an electric device in your home.

Activity 12

Observe Like a Scientist

Fossils in Our Lives

Watch the video and **look** for issues with using fossil fuels.

Quick Code:
ca4492s

Video

Fossils in Our Lives

Talk Together

Now, talk together about how people can benefit from conserving energy.

CCC **Energy and Matter**

Activity 13

Analyze Like a Scientist

Quick Code:
ca4493s

Conserving Fossil Fuels

Read the text and **look** at the picture. As you read, **underline** the main idea of the passage and **highlight** ways to conserve fossil fuels.

Conserving Fossil Fuels

The supply of fossil fuels in Earth's crust is limited. Since they are nonrenewable, this means that they will eventually run out. Using less fossil fuel to meet our needs is the best way to conserve these resources. There are many ways to conserve fossil fuels, such as walking or biking instead of driving, taking the bus or other public transportation, and turning off the lights when you are not in a room.

Replacing fossil fuel energy with renewable sources of energy such as solar power and wind power can greatly reduce the use of fossil fuels. The main drawback has been that it costs more to produce energy from renewable sources than from fossil fuels.

Bike Parked at Renting Station

Activity 14

Observe Like a Scientist

Quick Code:
ca4494s

Value of Renewable Resources

Complete the interactive. **Look** for how renewable sources of energy can be harnessed. **Complete** the table about renewable sources of energy.

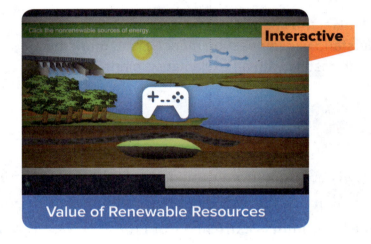

Value of Renewable Resources

Element	Renewable Source of Energy Used	Description (Working or Uses)
Dam		
Solar collectors		
Windmills		

SEP Engaging in Argument from Evidence

CCC Energy and Matter

Activity 15

Analyze Like a Scientist

Quick Code:
ca4495s

Nuclear Energy

Read the text and **look** at the image. As you read, **mark** the areas that you don't understand with a question mark (?), and **mark** the areas that you find interesting with an exclamation mark (!).

Nuclear Energy

Some electricity is obtained from radioactive rocks that are mined from Earth. These rocks make thermal energy and dangerous radiation. A substance called uranium is extracted from these rocks. Uranium can be used in nuclear power plants. The uranium is used to heat water to make steam. The steam is used to drive turbines and generate electricity. Uranium is not strictly a fuel because it does not burn, but uranium is used up. It is nonrenewable.

Scientists are trying to copy the way the sun makes energy. This is proving very difficult. They haven't succeeded yet in using this type of nuclear energy to make electricity.

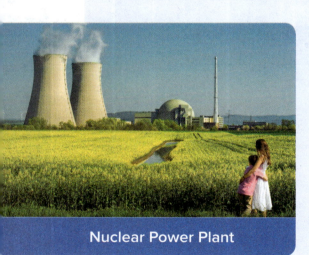

Nuclear Power Plant

SEP   **Obtaining, Evaluating, and Communicating Information**

Activity 16

Evaluate Like a Scientist

Quick Code:
ca4496s

Using Fuels

Write the fuels in the correct category.

charcoal   wood   gasoline

kerosene   ethanol   vegetable oil

Renewable	Nonrenewable

© Discovery Education | www.discoveryeducation.com ● Image: Mathias Genterczewsky / EyeEm / Getty Images

Activity 17

Record Evidence Like a Scientist

Fuels and Road Trips

Quick Code:
ca4497s

Now that you have learned about fuel, look again at the image Fuels and Road Trips. If you need to, go back and read the text. You first saw this in Wonder.

Let's Investigate Fuels and Road Trips

Talk Together

How can you describe fuels and road trips now? How is your explanation different from before?

Look at the Can You Explain? question. You first read this question at the beginning of the lesson.

Can You Explain?

How are fossil fuels formed and used?

Now, you will use your new ideas about fuels to answer a question.

1. **Choose** a question. You can use the Can You Explain? question or one of your own. You can also use one of the questions that you wrote at the beginning of the lesson.

My Question

2. Then use the sentence starters that follow to answer the question.

   To plan your scientific explanation, first **write** your claim. Your claim is a one-sentence answer to the question you investigated. It answers: What can you conclude? It should not start with yes or no.

   My claim:

   _____

   _____

   _____

   _____

   _____

   Data should support your claim. Leave out information that does not support your claim. **List** data that supports your claim.

Data 1

Data 2

Finally, **explain** your reasoning. Reasoning ties together the claim and the evidence. Reasoning shows how or why the data count as evidence to support the claim.

Evidence	Reasoning That Supports Claim

Now, write your scientific explanation.

Fossil fuels are formed by...

_____

_____

_____

_____

_____

_____

_____

_____

_____

_____

_____

_____

_____

_____

SEP   **Constructing Explanations and Designing Solutions**

Discovery
EDUCATION

 **in Action**

Quick Code:
ca4498s

Activity 18
Analyze Like a Scientist

Oil Drillers and ROVs

Read the text and **complete** the activities that follow.

Oil Drillers and ROVs

Technology and Safety

Oil is buried deep within sediments on the ocean floor. How do we get the oil? That job goes to oil drillers. Oil drillers work on oil rigs and use special equipment to extract the oil. Once oil drillers identify a specific location that has oil, they use long drilling pipes to make a hole in the ocean floor. The strong drill can cut through layers of hard rock if necessary. When the drill reaches the oil, it is replaced with a pump and a machine that pulls the oil upward.

Some oil deposits are found at the bottom of the ocean. They are too deep for humans to dive down to investigate them. Instead, robots take over. Robots can be remotely controlled to adjust parts of the drilling pipes. They carry video cameras that send pictures to a computer on the oil rig. Robots and other technologies help the oil drillers keep the oil rig, pipes, and pumps working properly.

SEP **Constructing Explanations and Designing Solutions**

Oil drillers have a dangerous job. Explosions, fires, oil leaks and spills, and exposure to dangerous chemicals are all possible. Thus, the use of robots and other specialized equipment that can replace human activity is very beneficial.

What other ways can technology assist with keeping oil drillers safe?

Working Like a Scientist

In April 2010, the Deepwater Horizon oil rig on the Gulf of Mexico exploded and sank. This accident caused the deaths of eleven people working on the rig. It also caused 3.19 million barrels (over 130 million gallons) of oil to spill into the ocean. Normally, oil and natural gas from the ocean floor flow up through the pipes and onto the rig. Oil and natural gas burn easily and can cause an explosion when they come in contact with a flame or certain chemicals. It is hard to stop the burning of oil and natural gas once it starts.

Suppose you are an oil rig safety specialist assigned to make the safety rules on an oil rig. **List** three of your safety rules that would help prevent accidents and/or explosions.

_____

_____

_____

_____

_____

Oil Spills

The pipes that oil drillers use to get oil and natural gas from the ocean floor have safety features. Sometimes, these safety features fail, and part of the pipe explodes. The oil and natural gas leak out into the ocean and even into the atmosphere. These pollutants affect living things in the ocean as well as the living things that feed upon these marine organisms.

Oil Rig Spill

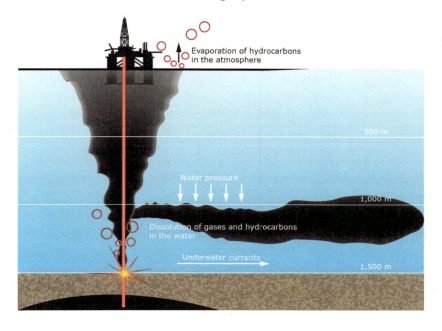

Review the illustration of the oil spill. **Describe** two ways to prevent spills like this from happening.

_____

_____

_____

_____

Activity 19

Evaluate Like a Scientist

Quick Code:
ca4499s

Review: About Fuels

Think about what you have read and seen in this lesson. **Write** down some core ideas you have learned. **Review** your notes with a partner. Your teacher may also have you take a practice test.

_____

_____

_____

_____

_____

_____

_____

_____

_____

_____

_____

_____

_____

_____

SEP Obtaining, Evaluating, and Communicating Information

Talk Together

Think about what you saw in Get Started. Use your new ideas to discuss fuels, how they are formed and used, and how we can use them wisely.

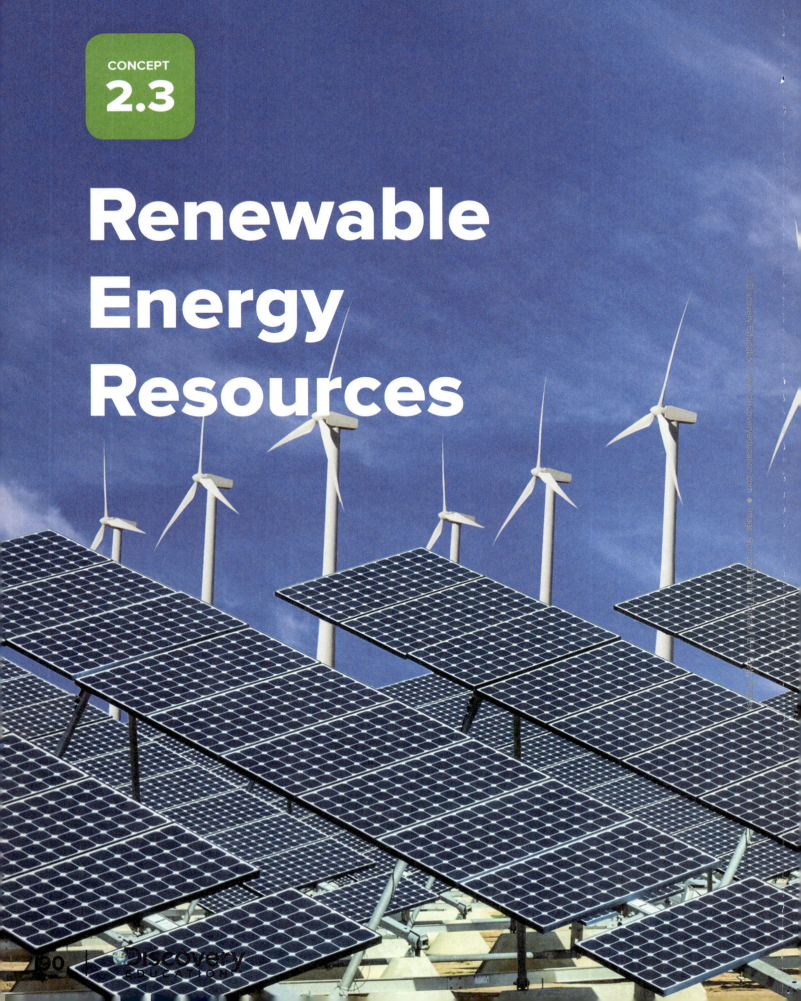

Renewable Energy Resources

Discovery
EDUCATION

Student Objectives

By the end of this lesson:

- [ ] I can apply scientific ideas to design, test, and refine devices that convert energy from one form to another.

- [ ] I can construct explanations for the use of solar radiation, wind, and falling water in the generation of electricity.

- [ ] I can develop models based on observations and evidence that energy is transferred from place to place by light, heat, and electricity.

Key Vocabulary

- [ ] geothermal
- [ ] heat
- [ ] light
- [ ] radiation
- [ ] renewable resource
- [ ] resource
- [ ] solar energy

Quick Code:
ca4500s

Activity 1

Can You Explain?

What are the different ways we can use renewable energy to generate electricity?

_____

_____

_____

_____

_____

_____

_____

_____

_____

_____

_____

Quick Code:
ca4502s

Activity 2

Ask Questions Like a Scientist

Quick Code:
ca4503s

Animal Energy

Read the text, and **look** at the pictures. Then, **complete** the activity that follows.

Animal Energy

Imagine you were born 400 years ago. This was before the invention of engines. People used machines before that date, but they were not powered by engines. There was no electricity. How were the machines powered? Some were powered by animals, such as using horses to pull carts. Humans did a lot of the work. They used simple machines like wheelbarrows and pulleys.

Life was hard. A few machines were powered by wind and water. What were these like? Ships were powered by the wind. The wind was also used to turn sails on windmills. These windmills pumped water and ground corn. There were other types of mills. These were water mills. They used the energy from flowing water to make them turn. Can you think of some of the advantages these early mills had? What about disadvantages?

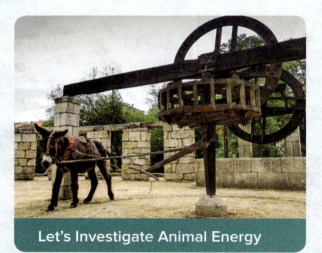

Let's Investigate Animal Energy

One of the most common jobs of windmills and water mills was to crush grain to make flour. This milling took place at the mill. The energy had to be used on site. There was no electricity system to carry the energy over a long distance. And what happened if the wind didn't blow or the river didn't contain enough water to power the mill? Old windmills and water mills are examples of how wind and water can be used to provide useful energy. Can you think of any modern ways that wind and water can be used to make energy?

Windmill

Write other questions you would like to investigate about the devices you use in your daily life and the energy they use.

SEP **Asking Questions and Defining Problems**

CCC **Energy and Matter**

Activity 3

Observe Like a Scientist

Under Sail

Look at the picture and **answer** the questions that follow.

Quick Code:
ca4504s

Under Sail

Where did these ships get their energy from?

_____

_____

What would happen if this form of energy became unavailable?

_____

_____

© Discovery Education | www.discoveryeducation.com ● Image: (a) Ron and Patty Thomas / E+ / Getty Images. (b) Pixabay

Activity 4

Observe Like a Scientist

Water Wheel

Quick Code:
ca4505s

Look at the picture and **answer** the questions that follow.

Water Wheel

What would the water wheel have been able to do for people?

_____

_____

Why don't we have many water wheels like this today?

_____

_____

Observe Like a Scientist

Travel without Engines

Look at the picture and **answer** the question that follows.

Travel without Engines

What is innovative about this cart, and how did it make travel easier?

_____

_____

_____

Activity 6

Evaluate Like a Scientist

What Do You Already Know About Renewable Energy Resources?

Sources of Energy

Draw a line to **match** each item with the source of its power.

flashlight	gasoline
solar-powered calculator	natural gas or electricity
cooktop burner	chemicals
car engine	sunlight
plant leaf	wood
baseball flying toward the catcher	light
campfire	pitcher's arm
firecracker	battery

Renewable or Not?

Circle the renewable sources of energy.

hydropower

wind

coal

sunlight

natural gas

What Are Some of the Different Ways We Use Energy from the Sun?

Activity 7

Observe Like a Scientist

The Sun

© Discovery Education | www.discoveryeducation.com ● Image (a) Ron and Patty Thomas / E+ / Getty Images, (b) Jiiladda Chaiyajina / Shutterstock.com, (c) Icon made by Freepik from www.flaticon.com

Quick Code:
ca4508s

Watch the video and **think about** how the sun produces light and heat.

Video

The Sun

Talk Together

Now, talk together about the conditions inside the sun.

SEP Obtaining, Evaluating, and Communicating Information
CCC Energy and Matter

Using Energy from the Sun

Quick Code:
ca4509s

Read the text. As you read, **draw** an image for each of the first two paragraphs.

Using Energy from the Sun

You see and feel sunlight. Even at night when you cannot see the sun in the sky, you feel the warmth of the sun's energy absorbed by the atmosphere. The land and water on Earth's surface also absorb the sun's energy, causing their temperatures to increase. If you look at the moon in the night sky, you see the **light** of the sun reflected off the moon's surface. Sunlight is radiant energy or **radiation**.

Life would not be possible without sunlight. Almost every plant and animal depends on energy from the sun to survive. Plants feed on sunlight through photosynthesis. Animals, including humans, feed on plants. In this way, sunlight supports all life on Earth.

SEP Obtaining, Evaluating, and Communicating Information
CCC Energy and Matter

Using Energy from the Sun *cont'd*

Energy received from the sun is called **solar energy**. We can use solar energy directly as a thermal energy source. Greenhouses capture the energy from the sun. This energy warms up the inside of the greenhouse. This helps farmers grow crops that would normally only grow in warm climates. Houses, too, can be built in a way that enables the energy from the sun to warm them up. This is usually done by placing large windows on the wall that faces the sun for the longest part of the day. Solar energy can also be used for cooking.

Time for Lunch!

In the spaces provided, **draw** an image for each of the first two paragraphs.

Paragraph 1

Paragraph 2

Activity 9

Observe Like a Scientist

What Is Light?

Quick Code:
ca4510s

Watch the video and **look** for how light travels from one place to another.

Video

What Is Light?

Talk Together

Now, talk together about how light energy from the sun reaches us here on Earth.

SEP   Obtaining, Evaluating, and Communicating Information

CCC   Energy and Matter

Activity 10
Observe Like a Scientist

Quick Code:
ca4511s

Solar Energy

Look at the images and **answer** the question that follows.

Greenhouse

Solar Panels

How do these energy systems or devices work?

_____

_____

_____

_____

_____

SEP   **Engaging in Argument from Evidence**

CCC   **Energy and Matter**

Activity 11

Observe Like a Scientist

Sunlight

Quick Code:
ca4512s

Complete the interactive, and **answer** the questions that follow.

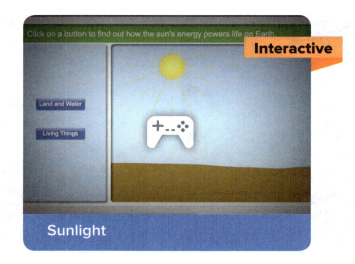

What does solar energy have to do with the water cycle?

_____

_____

_____

_____

_____

_____

_____

CCC **Energy and Matter**

Cheese comes from cows. People eat cheese to get energy. Explain where this energy originally came from.

_____

_____

_____

_____

_____

_____

Is it better to use sunlight or coal to make electricity? Why?

_____

_____

_____

_____

_____

_____

_____

Activity 12
Investigate Like a Scientist

Quick Code:
ca4513s

Hands-On Investigation: Make a Solar House

In this investigation, you will work in groups to design and construct a house that uses a solar cell as its primary energy source. You will test the solar panel at different angles to the light source. You will also test different places to mount the solar panel on your model. You will use the solar cell to provide lighting for your model house.

Make a Prediction

Why is the angle of the solar cell important?

_____

_____

_____

What is an example of an inefficient place to mount a solar panel?

_____

_____

_____

SEP	Developing and Using Models
CCC	Energy and Matter
CCC	Structure and Function

What materials do you need?

- Shoebox with lid

- Electrical tape

- Scissors

- 4 insulated wires with alligator clips

- Two 6-volt mini-bulbs with socket base

- Mini solar panels with alligator clips

- Encapsulated solar cell array

- Optical illusion disc

What Will You Do?

1. Create a simple circuit and light your light bulb using your solar panel.

2. Experiment by holding your solar panels at different angles to your light source. What effect did the angle have on lighting the bulb?

3. With your group, build your home.

4. Test the panel at different angles to the light source, and also test different places to mount the panels on your model.

5. Consider the variables shown in the following table. Record your observations in the table.

Variable	Observation	How does it model solar energy?
The angle of the tilt of the solar panels		
The length of time necessary for exposure to the sunlight to light the bulb		
The amount of sunlight necessary to turn on the light bulb		

Discovery
EDUCATION

Think About the Activity

Did your original plan for the solar house work? What did you find most challenging?

_____

_____

_____

How did the angle and location of the solar panels affect the efficiency of your design?

_____

_____

_____

Compare your group's house to another group's. How are they similar? How are they different? How did these differences affect your results?

_____

_____

_____

_____

Activity 13
Analyze Like a Scientist

Quick Code:
ca4514s

Solar Energy

Read the text. As you read, **underline** evidence in the text that energy is transformed from one form to another.

Solar Energy

Solar energy can also be used to ==heat== water. Panels made of black pipes can be placed on the roofs of house. As water passes through the pipes, it heats up. It can then be stored in a hot water tank for use.

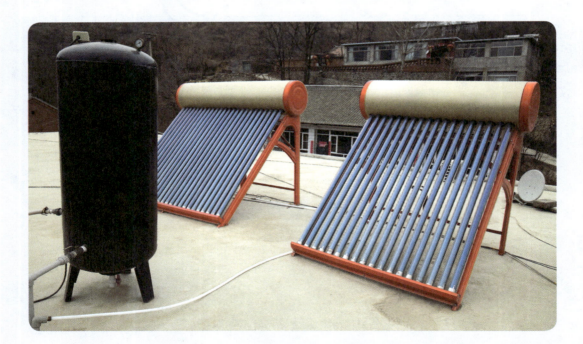

SEP   Constructing Explanations and Designing Solutions

CCC   Energy and Matter

Most solar panels are used to generate electricity. Solar panels that generate electricity are made of many small solar cells. These cells catch the radiant energy of the sun and turn it directly into electricity. This is called solar power. The electricity charges batteries, and runs lights and other devices. Solar-cell calculators run on batteries powered by small solar cells. Houses and other buildings may use electricity made from rooftop solar panels.

The solar cells that are used in solar panels can be used to drive other devices. Your yard may contain solar lights to light up the driveway at night. People build and race cars that use solar energy. They have even made a plane with solar cells on its wings. It uses the electricity the solar cells make to power its propellers.

Activity 14
Evaluate Like a Scientist

Quick Code:
ca4515s

How Do We Use Solar Energy?

Write "Some," "No," or "All" to best complete the sentences.

x _____No_____ solar panels use the heat from the sun to heat water or air.

↓ _____Some_____ solar panels use the light from the sun to produce electricity.

+ _____All_____ solar panels directly create electricity using the sun's heat.

↓ _____All_____ solar panels work only during daylight hours.

How Can We Capture the Wind to Provide Useful Energy?

Activity 15
Analyze Like a Scientist

Harness the Wind

Read the text. Then, **complete** the activity that follows.

Quick Code:
ca4516s

Harness the Wind

As the sun warms Earth, it warms the air. Different parts of the world get different amounts of this solar energy. This causes the air to move and the wind to blow. We can use the energy in the wind to turn windmills.

Where have you seen windmills? Maybe you have seen one on an old farm where it is used to pump water from wells. Have you seen ones that are more modern? Perhaps you have passed a wind farm. Wind farms have many windmills. These windmills are used to make or generate electrical energy. For this reason, they are called wind generators or wind turbines. The electricity from wind turbines is carried by big wires to places where it is needed.

The wind also makes waves on the surface of the sea. These waves contain lots of energy. Wave energy can be used to generate electricity.

SEP Obtaining, Evaluating, and Communicating Information

CCC Energy and Matter

Draw an energy chain showing the inputs and outputs of a windmill in a wind farm.

Activity 16

Observe Like a Scientist

Wind Power

Watch the video and **look** for how wind turbines turn the kinetic energy of wind into electricity.

Green Revolution: Wind Power

Talk Together

Now, talk together about the locations you think are best for wind turbines.

| SEP | Obtaining, Evaluating, and Communicating Information |
| CCC | Energy and Matter |

How Can Energy from Falling Water Be Used to Generate Electricity?

Activity 17
Think Like a Scientist

Quick Code:
ca4518s

Modeling a Turbine Generator

In this investigation, you will use a pinwheel and siphon to model a spinning turbine in a hydroelectric dam.

What materials do you need?

- 2 containers of at least gallon capacity, one of which is filled with water

- Flexible tube approximately 1 meter long

- Pinwheel
  - Manila folder
  - Scissors
  - Tape
  - Pencil
- Plastic cup or beakers

What Will You Do?

1. Use the materials to model a turbine generator.

2. When the water runs out, one person in your group can use a cup to transfer water to the upper container.

Discovery
EDUCATION

Think About the Activity

How did the pinwheel and siphon model a hydroelectric power station?
Describe or diagram your model.

Describe how you changed your model so it ran on renewable energy.

_____

_____

_____

SEP	Developing and Using Models
CCC	Energy and Matter
CCC	Structure and Function

Which alternative energy resources come from forms of mechanical energy?

_____

_____

How can mechanical energy be used to generate electricity?

_____

_____

_____

Analyze Like a Scientist

Quick Code:
ca4519s

Falling Water

Read the text. As you read, **record** similarities and differences between using water and using wind to generate electricity in the graphic organizer that follows.

Falling Water

The sun's energy warms Earth's water and causes it to evaporate. The evaporated water condenses to form clouds. Eventually those clouds fall as rain or snow. Rain and snow provide water for rivers. Rivers have a lot of energy we can use. Indirectly, the energy from running water comes from the sun.

Water Cycle

Falling Water *cont'd*

Rivers run downhill. As they run, they change gravitational **potential energy** to kinetic energy. We can use this energy to turn a water mill. We can also use it to generate electricity. Some big dams contain turbines. The water builds up behind these dams. When the water is let out, it passes through turbines in the dam. The water makes the turbines turn. The turbines and generators in the dam generate electricity. The electricity can be sent along wires to cities where it is needed. This type of electricity is called hydroelectricity.

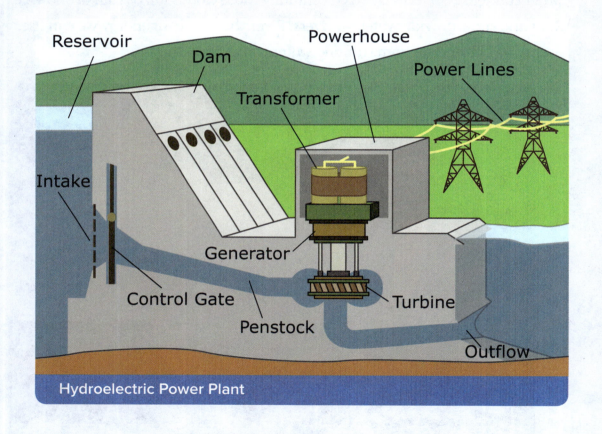

Hydroelectric Power Plant

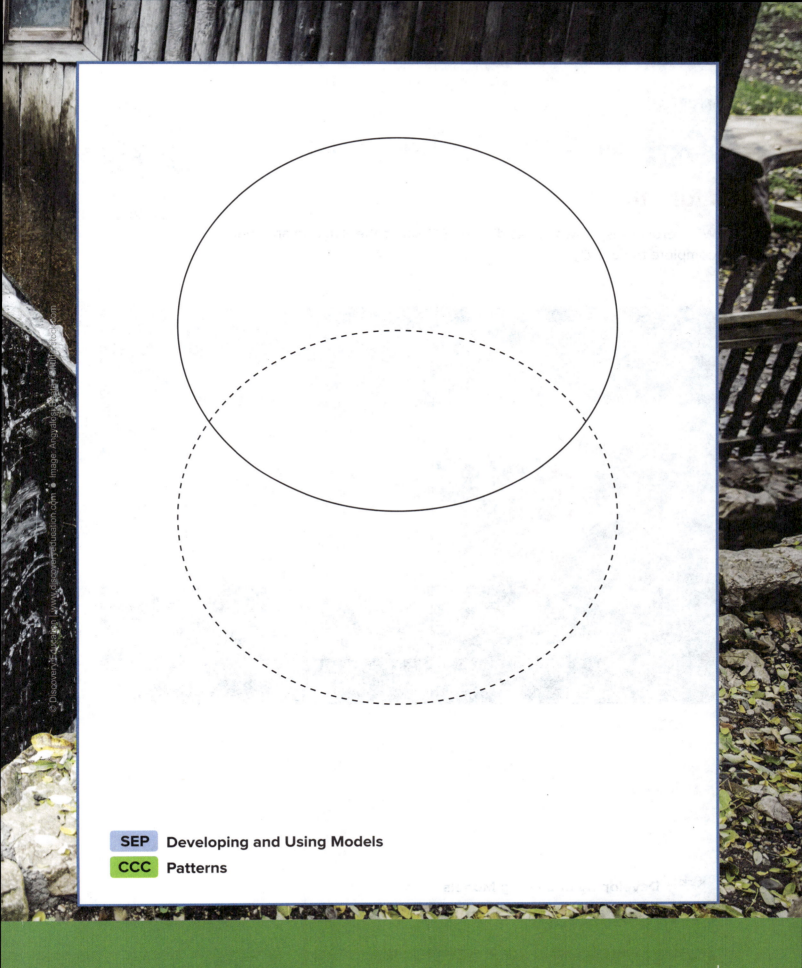

SEP Developing and Using Models

CCC Patterns

Activity 19

Evaluate Like a Scientist

Light the Bulb

What process is shown in this diagram? **Discuss** the diagram and then complete the activity.

Quick Code:
ca4520s

© Discovery Education | www.discoveryeducation.com • Image: (a) Ron and Patty Thomas / E+ / Getty Images. (b) Paul Fuqua

SEP Developing and Using Models

CCC Energy and Matter

A student is diagramming how hydroelectric power can provide the energy for light bulbs in her classroom. The diagram starts with the water cycle and how it moves water across Earth, and it ends with the bulbs in the classroom. Number the terms to show the correct sequence.

Kinetic energy _____

Electrical energy _____

Gravitational potential energy _____

Solar energy _____

Light and heat energy _____

Activity 20

Record Evidence Like a Scientist

Quick Code:
ca4521s

Animal Energy

Now that you have learned about animal energy, look again at the image Animal Energy. You first saw this in Wonder.

Let's Investigate Animal Energy

Talk Together

How can you describe animal energy now? How is your explanation different from before?

SEP   Constructing Explanations and Designing Solutions

Look at the Can You Explain? question. You first read this question at the beginning of the lesson.

Can You Explain?

What are the different ways we can use renewable energy to generate electricity?

Now, you will use your new ideas about renewable energy resources to answer a question.

1. **Choose** a question. You can use the Can You Explain? question or one of your own. You can also use one of the questions that you wrote at the beginning of the lesson.

My Question

2. Then, use the graphic organizers on the next pages to help you answer the question.

To plan your scientific explanation, first **write** your claim. You claim is a one-sentence answer to the question you investigated. It answers: What can you conclude? It should not start with yes or no.

My claim:

_____

_____

_____

_____

_____

_____

_____

_____

_____

_____

_____

_____

Finally, **explain** your reasoning. Reasoning ties together the claim and the evidence. Reasoning shows how or why the data count as evidence to support the claim.

Evidence	Reasoning That Supports Claim

What are the different ways we can use renewable energy to generate electricity?

Now, write your scientific explanation.

 **in Action**

Activity 21

Analyze Like a Scientist

The Green Solar Revolution

Read the text, **watch** the video, and **record** some of the different types of research that is now taking place on solar and plant cells.

The Green Solar Revolution

Fossil fuels are made from the remains of dead plants and animals. Examples of fossil fuels are oil, natural gas, and coal. For centuries, people around the world have burned fossil fuels for energy. However, there are many disadvantages to using fossil fuels. One is that they create large amounts of pollution. Another is that they are a nonrenewable resource. One day, we will run out of fossil fuels.

This is why teams of scientists have been working for decades to find clean and renewable forms of energy. Solar energy is a good source of clean, renewable energy. Researchers around the world have found ways to collect, concentrate, and store that energy. These scientists work as teams in laboratories and in the field. Some of these researchers work at universities. Others work for governments—trying to solve energy and pollution problems for their country. And still others work for private companies that are trying to design and develop solar technologies that can meet the daily needs of both people and businesses.

The Green Solar Revolution *cont'd*

Let's learn about the work of some solar energy researchers at Arizona State University.

The work of researchers is making solar energy much more available to us. Solar energy now comes in portable forms. Some small, personal devices use solar energy. For example, there are backpacks with built-in solar panels. People can use them to charge their cell phones, tablets, and laptops while they are walking or biking from place to place. Researchers are also designing smaller, cheaper, and more flexible solar cells. In some areas, houses and even large office buildings are powered by solar energy.

Video

Green Revolution: Solar Power

Record some of the different types of research that is now taking place on solar and plant cells.

_____

_____

_____

_____

_____

Discovery EDUCATION

Solar and Plant Cells

Think about photovoltaic solar cells and plant cells, and what they can do. How are photovoltaic solar cells and plant cells similar? How are they different?

_____

_____

_____

_____

_____

_____

| SEP | Energy and Matter |
| CCC | Structure and Function |

Activity 22

Evaluate Like a Scientist

Quick Code:
ca4523s

Review: Renewable Energy Resources

Think about what you have read and seen in this lesson. Write down some core ideas you have learned. Review your notes with a partner. Your teacher may also have you take a practice test.

_____

_____

_____

_____

_____

_____

Talk Together

Think about what you saw in Get Started. Use your new ideas to discuss renewable energy sources, how they work, how they are similar, and how they are different.

SEP Obtaining, Evaluating, and Communicating Information

Discovery EDUCATION

Energy and the Environment

Discovery
EDUCATION

Student Objectives

By the end of this lesson:

- [ ] I can argue from evidence that extracting and using energy and fuels affects the local and global environment.

- [ ] I can develop models that demonstrate how energy is transferred from place to place by light, heat, and electricity.

- [ ] I can design and compare solutions that reduce the amount of electricity we use and evaluate how effective they are.

Key Vocabulary

- [ ] air
- [ ] ecosystem
- [ ] environment
- [ ] pollution

Quick Code:
ca4524s

Activity 1

Can You Explain?

How can you reduce the impact of energy production on the environment?

_____

_____

_____

_____

_____

_____

_____

_____

_____

_____

Quick Code:
ca4526s

Activity 2
Ask Questions Like a Scientist

Quick Code:
ca4527s

Smog Pollution

Read the text and **look** at the image. Then, **complete** the activity that follows.

Smog Pollution

Heat waves are no fun. Phew! It is too hot to do anything except perhaps go for a swim or spray your friends with a garden hose. A more serious problem with heat waves is **air** quality.

Let's Investigate Smog Pollution

Last year in the city where Justin lives, they had a warning—unhealthy to very unhealthy air. This unhealthy air was because of a gas called ozone. Ozone is a form of pollution. This gas is formed when gases put out by cars and trucks bake in the heat and sunlight. If these get trapped near the ground, then breathing can become difficult. Ozone pollution is always bad during heat waves. Justin's friend has asthma. When the air gets very bad, she starts to wheeze and must use her inhaler. Justin also finds it difficult to breathe. Even his dog gets sick. When pollution gets that bad, they both play inside.

They call this bad air smog. It is particularly bad in their city because they live in a valley. The valley sides trap the smog and stop it from being blown away. It would be nice if they didn't have smog. But how can the city reduce it, or better still, get rid of it entirely?

Discuss with Your Class

What do you think a heat wave is? Do you think they are common? Can you remember what you did during the last heat wave? Did you notice anything about the color and smell of the air during the heat wave? What questions do you have about heat waves?

_____

_____

_____

SEP **Asking Questions and Defining Problems**

CCC **Cause and Effect**

Activity 3
Observe Like a Scientist

Quick Code:
ca4528s

Smog over LA

Look at the photograph of smog over the city of Los Angeles.
Discuss the question that follows.

Smog over LA

Talk Together

What are some impacts of smog on humans and the environment?

Activity 4
Observe Like a Scientist

Quick Code:
ca4529s

Smog in China

Watch the video and **look** for ways that people respond to high pollution days.

Smog in China

Talk Together

Now, talk together about ways that smog could be reduced.

Activity 5
Observe Like a Scientist

One Idea for Reducing Smog

Watch the video and **look** for ways that people are affected by smog.

Quick Code:
ca4530s

Reducing Smog

Talk Together

Now, talk together about additional questions you have about the causes and effects of smog.

© Discovery Education | www.discoveryeducation.com ● Images: (a) yangphoto / E+ / Getty Images. (b) designbydx / Shutterstock.com. (c) Icon made by Freepik from www.flaticon.com

Activity 6

Evaluate Like a Scientist

Quick Code:
ca4531s

What Do You Already Know About Energy and the Environment?

Breathe Easy

Highlight the actions that can help reduce air pollution.

Riding a bike instead of driving

Installing carbon scrubbers at power stations

Closing nuclear power plants

Reducing power outages

Using new materials instead of reusing materials

Planting trees

CCC **Cause and Effect**

Energy Impacts

Draw a line to **match** each energy source to its potential environmental impact.

wind	toxic manufacturing byproducts
coal	fracking damage
solar	bird deaths
oil	mountaintop removal
natural gas	flooding valleys
hydroelectric	pipeline/tanker spills

How Does Removing Fossil Fuels Impact the Environment?

Activity 7

Analyze Like a Scientist

Impact of Fossil Fuels on the Environment

Quick Code:
ca4532s

Read the text and **underline** evidence that using fossil fuels affects the environment.

Impact of Fossil Fuels on the Environment

Earth has a limited supply of fossil fuels, but a lot of coal, gas, and oil reserves still remain underground. Many scientists think we should reduce our use of these sources of energy. Some suggest we should stop using them altogether. They suggest we switch to renewable energy sources. Why do they think this?

SEP Engaging in Argument from Evidence

CCC Energy and Matter

CCC Cause and Effect

Fossil fuels must be extracted from the ground. This process damages the environment. Coal is mined. Usually, this forms big pits. Sometimes, whole mountain tops are removed. The waste from mines must be dumped somewhere. Often, it is dumped in nearby valleys. This pollutes the water supply. The cost of cleaning up coal mines is very high.

Drilling and extracting oil and gas also disrupts **ecosystems**. A lot of oil is removed from wild remote areas. Rigs, roads, and pipelines are needed to extract these fuels, disrupting wildlife. Accidents occur, and oil is spilled. A lot of oil extraction occurs at sea. If oil is spilled into the sea, it pollutes the water. Oil kills sea life and damages beaches.

Mountaintop Removal for Coal Mining

Activity 8
Observe Like a Scientist

Stopping the Spread

Watch the video and **look** for how and why people were trying to contain oil from the spill.

Stopping the Spread

Talk Together

Now, talk together about why it is important to stop oil from spreading.

CCC   **Cause and Effect**

Activity 9

Investigate Like a Scientist

Quick Code:
ca4534s

Hands-On Investigation: Oil Spill Cleanup

In this investigation, you will create a model of an oil spill. Then, you will explore different methods used to clean and contain oil spills. Finally, you will build a model sorbent boom to explore the effectiveness of booms to contain oil spills.

Make a Prediction

What do you think will make oil spills difficult to clean up?

_____

_____

_____

_____

_____

What do you predict will be some good ways of cleaning up oil spills?

_____

_____

SEP	Developing and Using Models
SEP	Constructing Explanations and Designing Solutions
CCC	Structure and Function

What materials do you need?

- 1 9-inch aluminum pie plate

- 1 stone that fills about 20 percent of the pie plate

- 1 small plastic funnel

- One 20 mL graduated cylinder and stand

- 2 bird feathers

- 2 cotton washcloths

- 1 square of cotton muslin

- 1 small squeeze bottle of dishwashing soap

- 1 roll of masking tape

- 2 twist ties

- Plastic spoon

- Pitchers of water

- Squeeze bottles containing vegetable oil mixed with cocoa powder

- Paper towels and garbage bags for cleanup

What Will You Do?

1. Place stones and water in your pie plate.

2. Spill the vegetable oil into the water.

3. Observe how the oil moves and behaves, then write down your observations.

4. Place a feather in the water for 30 seconds.

5. Try skimming the oil off the water with a spoon.

6. Try to soak up the oil with one of your cloths.

7. Squeeze a few drops of dishwashing soap into your pan.

8. Clean your pan and refill it with a new "oil spill."

9. Build a model boom by wrapping the muslin square around a rolled-up cloth.

10. Try to keep the oil from reaching the rocks by using the boom.

Think About the Activity

Which method is most effective for cleaning spilled oil from water? Which method is most effective for cleaning oil from solid objects (rocks and bird feathers)?

_____

_____

_____

_____

_____

_____

Are booms an effective way to contain an oil spill? Why or why not?

_____

_____

_____

_____

What advice would you give to an oil company about containing and cleaning a spill?

_____

_____

_____

_____

How Can Burning Fossil Fuels Damage the Environment?

Activity 10

Observe Like a Scientist

Quick Code:
ca4535s

Contaminated Air and Contaminated Water

Watch the video and **look** for how chemicals in the air can end up in bodies of water.

Contaminated Air and Water

Talk Together

Now, talk together about how air pollution can cause harm to both aquatic and terrestrial ecosystems.

CCC Cause and Effect   **CCC** Energy and Matter

Activity 11
Observe Like a Scientist

Quick Code:
ca4536s

Big City Environmental Concerns

Watch the video and **look** for reasons why big cities have air pollution problems.

Big City Environmental Concerns

Talk Together

Now, talk together about the sources of air pollution in big cities.

CCC Cause and Effect

CCC Energy and Matter

DISCOVERY EDUCATION

Activity 12

Analyze Like a Scientist

Burning Fossil Fuels and Pollution

Read the text. As you read, **complete** the graphic organizer at the end of the passage to show how burning fossil fuels affects the environment.

Burning Fossil Fuels and Pollution

When fuels are burned, they produce different substances. Most of these substances go into the air. For example, when fossil fuels are burned, they give off carbon dioxide and water vapor. Carbon dioxide causes the planet to heat up. Scientists are worried that as Earth warms up, warming will cause harmful changes to life on Earth.

Coal-Fired Power Plant

Fossil fuels contain other impurities that make more harmful gases. When coal and oil products are burned, they make gases that dissolve in rain. These gases make the rain acidic. Acid rain can be very harmful. It can kill trees. It can change the chemistry of lakes and kill fish. It can change the chemistry of soil. Acid rain can dissolve some rocks, including some used for building.

CCC **Cause and Effect**

CCC **Energy and Matter**

Burning Fossil Fuels and Pollution *cont'd*

Smoke and gases from stacks and car tailpipes can combine to form smog. Smog can settle close to the ground and make it difficult to breathe.

Coal contains dangerous substances called heavy metals. One of these is mercury. When coal is burned, small amounts of mercury are released into the air. This mercury eventually settles over the land. It can poison fish and other wildlife.

Some of the pollution caused by fossils fuels can be reduced by removing the acid gases and mercury before they get into the air. Modern power plants have devices called scrubbers that remove some of these substances. Cars are fitted with catalytic converters on their tail pipes to remove some of the pollution.

Laws exist that regulate the pollutants that enter the air. These laws have saved hundreds of thousands of lives over the last few decades. Even so, each year in the United States, people die because of air pollution.

Topic: _____

Cause	Event	Effect

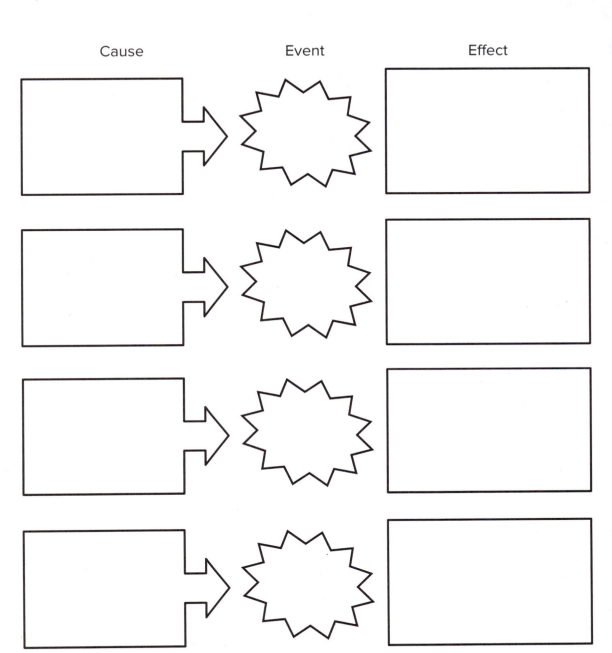

Activity 13

Investigate Like a Scientist

Quick Code:
ca4538s

Hands-On Investigation: Acid Rain

In this investigation, you will explore the effects of acid rain

Make a Prediction

Have you ever observed the effects of acid on organic matter?

_____

_____

_____

_____

_____

What effect do you think the vinegar will have on the chalk?

_____

_____

_____

_____

_____

SEP   Developing and Using Models

SEP   Planning and Carrying Out Investigations

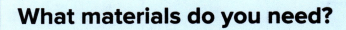

What materials do you need?

- 8 oz cup of distilled water
- 8 oz cup of vinegar
- Two 12 oz glass jars or beakers
- Masking tape and markers

- 2 strips of litmus paper
- 2 pieces of chalk
- 2 green leaves
- 2 pairs of latex gloves

What Will You Do?

1. With your partner, place a piece of chalk and a green leaf in both your "Distilled Water" and "Vinegar" containers.

2. The next day, observe the changes to the chalk and leaf in the vinegar container and compare them to the chalk and leaf in the water container. Record your observations.

3. A week later, observe the changes to the chalk and leaf in the vinegar container, then compare them to the chalk and leaf in the water container. Record your observations.

© Discovery Education | www.discoveryeducation.com

Think About the Activity

What effects of the vinegar did you observe after one day, compared to the effects of the distilled water?

_____

_____

_____

_____

_____

_____

What effects of the vinegar did you observe after a week, compared to the effects of the distilled water?

_____

_____

_____

_____

_____

_____

_____

_____

Discovery
EDUCATION

What conclusions can you draw from this experiment about the effects of acid rain on the environment?

_____

_____

_____

_____

_____

_____

_____

_____

Activity 14

Analyze Like a Scientist

Quick Code:
ca4539s

Problems, Solutions, and More Problems

Read the text. As you read, **add** to your cause and effect graphic organizer from the previous activity.

Problems, Solutions, and More Problems

In the 1800s, people began to need more energy than ever before. They needed energy to run factories, cars, trains, and ships.

Since then the demand for energy has continued to rise. More energy is needed to supply electricity to homes, schools, businesses, and factories. The problem has always been finding a way to get all this energy.

The solution was fossil fuels. Fossil fuels include coal, oil, and natural gas. Burning fossil fuels releases energy. People can use this energy to make things work. For example, people can burn coal in a power plant. They can use the energy from the coal to produce electricity. Then, they send the electricity through power lines to homes, schools, and factories.

CCC **Cause and Effect**

However, burning fossil fuels makes more than just energy. It also makes pollution. For example, burning coal creates a gas called sulfur dioxide. This gas rises in the air and mixes with water there. When this happens, it makes a chemical called sulfuric acid.

This coal will be burned in a power plant to produce electricity.

When it rains, sulfuric acid falls to Earth as acid rain. Acid rain can pollute the land and water. As a result, plants and animals can lose their habitats and even die.

There are several ways to reduce the pollution caused by burning coal. One way is to use coal that contains less sulfur. This helps to reduce acid rain.

Burning natural gas and oil produces a gas called carbon dioxide. Carbon dioxide can combine with water in the air to make carbonic acid. Like sulfuric acid, carbonic acid can also cause acid rain.

Carbon dioxide from burning fossil fuels can also cause another problem. Carbon dioxide gas can collect in the air. It forms a layer that traps heat on Earth. As a result, Earth's temperatures slowly rise. Rising temperatures on Earth is called global warming.

At the moment, the only solution to stop acid rain and global warming is to conserve energy. The less energy we use, the fewer fossil fuels we burn. The fewer fossil fuels we burn, the less carbon dioxide we put in the air.

Problems, Solutions, and More Problems *cont'd*

Conserving energy not only reduces pollution, it also conserves the supply of fossil fuels. Fossil fuels are nonrenewable. That means once fossil fuels are used, they cannot be used again. It takes millions of years to create more fossil fuels. Conserving fossil fuels makes them last longer and also keeps Earth cleaner.

Burning fossil fuels releases pollution into the environment.

Activity 15

Evaluate Like a Scientist

Quick Code:
ca4541s

Effects of Fossil Fuels

Write two examples of benefits and two examples of costs of using fossil fuels in the table below.

Benefit	Cost

Review the data in the table. Do the benefits outweigh the costs, or do the costs outweigh the benefits? Support your decision with information from the table.

_____

_____

_____

_____

_____

_____

CCC Cause and Effect   **CCC** Energy and Matter

What Are Some Negative Impacts of Renewable Energy?

Activity 16

Observe Like a Scientist

Is Making Fuel from Corn a Good Idea?

Quick Code:
ca4542s

Watch the video and **look** for problems associated with using corn to produce ethanol.

Video

Is Making Fuel from Corn a
Good Idea?

Talk Together

Now, talk together about the advantages and disadvantages of growing fuel.

SEP	Obtaining, Evaluating, and Communicating Information
CCC	**Energy and Matter**

Discovery EDUCATION

Activity 17
Evaluate Like a Scientist

Quick Code:
ca4543s

Effects of Deforestation

Circle all the effects of deforestation.

soil erosion

wildfire

insect outbreak

growth of new plants

disappearance of some animals

pollution

SEP **Constructing Explanations and Designing Solutions**

CCC **Cause and Effect**

Quick Code:
ca4544s

Activity 18
Analyze Like a Scientist

Negative Impacts of Renewable Energy

Read the text. As you read, **underline** possible negative impacts from these forms of renewable energy production.

Negative Impacts of Renewable Energy

What about solar energy? Once built, solar panels that generate electricity have very little impact. But making these panels damages the environment. They are made from substances that must be mined. Wind turbines also need to be made. Both also take up space, land that could be used for people, farming, or wildlife. Wind turbines kill a lot of birds. The best place for windfarms is on hills. Birds often follow hills on migration. Many fly into the turbines. So do bats. Millions of these animals die every year.

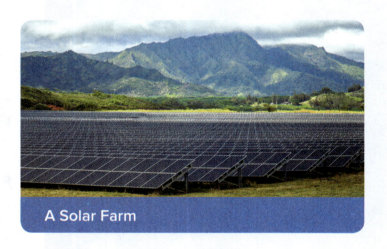

A Solar Farm

Big dams are built to make hydroelectric energy. These dams create massive lakes. The lakes drown land. Sometimes they drown complete cities and valuable archaeological sites. The dams block rivers. This can stop fish from swimming upstream. Some types of fish must swim upstream to lay their eggs. Building dams makes this impossible.

How Can We Reduce the Amount of Energy We Use?

Activity 19

Observe Like a Scientist

Quick Code:
ca4545s

How Big Is Your Footprint?

Complete Level 1 and Level 2 of the interactive. Once you have collected your data, **record** it in the tables below.

How Big Is Your Footprint?

SEP	Constructing Explanations and Designing Solutions
SEP	Developing and Using Models
CCC	Cause and Effect
CCC	Energy and Matter

Level 1

Test Number	Solar	Wind	Coal	Gas	Carbon Footprint Size	Total Cost: $ per week

Level 2

Test	Solar %	Wind %	Coal %	Gas %	Lights Replaced	Energy Cost per month	Carbon Footprint Size	CFL Cost	Total Cost per month

Activity 20

Evaluate Like a Scientist

Conservation Solutions

With a partner, **describe** the problem in each situation and **suggest** possible solutions.

Situation	Describe the Problem	Suggest a Solution
Darshana and Andy are neighbors. Darshana's dad drives her to a school picnic. Andy's mom drives him to the same picnic.		
Vivian turned on the computer and began typing an email. She went to get a snack, leaving the computer and the lights in her room on.		
Alex throws his plastic drink bottle in the trash.		

SEP   **Constructing Explanations and Designing Solutions**

Activity 21

Record Evidence Like a Scientist

Quick Code:
ca4547s

Smog Pollution

Now that you have learned about energy and the environment, look again at the image Smog Pollution. You first saw this in Wonder.

Let's Investigate Smog Pollution

Talk Together

How can you describe smog pollution now? How is your explanation different from before?

© Discovery Education | www.discoveryeducation.com ● Image: (a) yangphoto / E+ / Getty Images, (b) Andrew Gonis-EyeEm / EyeEm / Getty Images. (c) Icon made by Freepik from www.flaticon.com

Look at the Can You Explain? question. You first read this question at the beginning of the lesson.

Can You Explain?

How can you reduce the impact of energy production on the environment?

Now, you will use your new ideas about energy and the environment to answer a question.

1. **Choose** a question. You can use the Can You Explain? question or one of your own. You can also use one of the questions that you wrote at the beginning of the lesson.

My Question

2. Then, use the graphic organizers on the next pages to help you answer the question.

To plan your scientific explanation, first **write** your claim. Your claim is a one-sentence answer to the question you investigated. It answers: What can you conclude? It should not start with yes or no.

My claim:

_____

_____

_____

_____

_____

Data should support your claim. Leave out information that does not support your claim. **List** data that supports your claim.

Data 1

Data 2

Finally, **explain** your reasoning. Reasoning ties together the claim and the evidence. Reasoning shows how or why the data count as evidence to support the claim.

Evidence	Reasoning That Supports Claim

Now, write your scientific explanation.

_____

_____

_____

_____

_____

_____

_____

_____

_____

_____

_____

_____

_____

_____

_____

SEP   **Constructing Explanations and Designing Solutions**

S T E M in Action

Quick Code:
ca4548s

Activity 22

Analyze Like a Scientist

Careers and Renewable Resources

Read the text. As you read, **highlight** anything that surprises you. Then, **watch** the video and **look for** types of technology that harness wind and solar energy.

Careers and Renewable Resources

The United States gets 84 percent of its energy from fossil fuels. Fossil fuels include oil, coal, and natural gas. About 19 million barrels of oil are consumed each day in America. Each year, over one billion tons of coal are mined in the states. Most of this coal is used for electricity. The amount of natural gas used in the United States is expected to increase from 22.5 trillion cubic

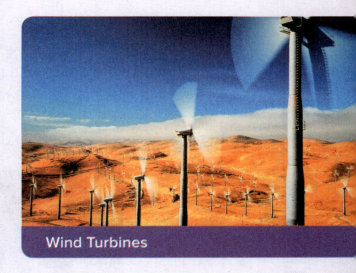

Wind Turbines

SEP	Obtaining, Evaluating, and Communicating Information
SEP	Constructing Explanations and Designing Solutions
CCC	Energy and Matter

Careers and Renewable Resources *cont'd*

feet (tcf) in 2009 to about 23.5 tcf in 2030. Eventually, these nonrenewable sources of energy will run out. Overuse of these energy sources can cause big problems! Burning fossil fuels promotes pollution, harmful smog, and global warming.

Switching to renewable sources of energy can help solve these problems. Making this change requires new inventions. There is good news. We have many sources of renewable energy and inventors and scientists who are developing related new technologies. Would you be interested in these types of careers? If so, you could play an important role in helping power our world with clean, renewable energy.

Wind energy is already in use around the country. Engineers design wind turbines that change wind energy to electrical energy. Wind techs install, maintain, and repair wind turbines. Scientists who study wind help determine the best places to place turbines. Large fields of turbines, like the one in California shown below, generate a lot of electricity. These wind turbines provide clean electricity to many homes and other buildings around the state.

Can you think of careers related to the use of solar energy?

Video

Wind and Solar Energy

Best Renewable Energy Choices

You have learned about different types of renewable energy resources. The best choice of renewable energy often depends on the situation. Match the situation described on the left with its corresponding best choice of renewable energy source.

Situation	Energy Type
fuel for indoor heating	coal
fuel for diesel vehicles	used vegetable oil
electricity for buildings in the southwest U.S. desert	gasoline
electricity for buildings along a coastline with constant winds	wood
	wind turbines
	nuclear power plant
	solar panels
	hydroelectric power

Activity 23

Evaluate Like a Scientist

Quick Code:
ca4549s

Review:
Energy and the Environment

Think about what you have read and seen in this lesson. Write down some core ideas you have learned. **Review** your notes with a partner. Your teacher may also have you take a practice test.

_____

_____

_____

_____

_____

_____

_____

_____

_____

_____

_____

_____

SEP Obtaining, Evaluating, and Communicating Information

Discovery
EDUCATION

Talk Together

Think about what you saw in Get Started. Use your new ideas to discuss renewable and nonrenewable resources and explain how the use of renewable resources can have a positive impact on the environment.

Solve Problems Like a Scientist

Unit Project: Dam Impacts

Quick Code:
ca4550s

In this project, you will analyze the effects that building the Hoover Dam had on the Colorado River, both upstream and downstream. **Read** the text and **complete** the activities that follow.

Hoover Dam

SEP Obtaining, Evaluating, and Communicating Information

CCC Structure and Function

Dam Impacts

Dams are built on rivers to hold back the flow of the water. They use the water's energy to power machines that make electricity. But dams also affect the environment that surrounds them. How do dams change the landscape? How does building a dam affect humans and wildlife that depend on the river?

In the middle of the vast Nevada desert, a giant lake sparkles. How odd it is to find a lake in the middle of a desert! Yet Lake Mead, which is more than 100 miles long, is the largest human-made reservoir in North America. It did not exist until the Hoover Dam was built in 1935.

In 1928, the Boulder Canyon Project Act was passed. It called for the construction of a dam to control the Colorado River. The water would be divided among California, Arizona, and Nevada. Black Canyon was chosen as the site of the dam, and thousands of people came from across the country to work on the project.

Positive or Negative?

Think about the results of the Hoover Dam. **Complete** the graphic organizer with the results listed.

- changing fish migration routes
- generating pollution from increased tourism
- producing hydroelectric power

- flooding the land upstream from the dam
- controlling the downstream river level
- providing a steady water supply

Positive	Negative

The Greatest Benefit

Choose the greatest benefit of building the Hoover Dam and **research** it. Then, **explain** why you chose that benefit as the best for humans, wildlife, and the landscape surrounding the Hoover Dam. Be sure to **list** all sources you used to research this answer.

_____

_____

_____

_____

_____

_____

Your Solutions

Choose one of the drawbacks of building the Hoover Dam and **research** possible solutions to that drawback. **Write** a description of why it is important to address this drawback and propose a solution. Be sure to **include** all your sources in the description.

_____

_____

_____

_____

_____

_____

Grade 4 Resources

- **Bubble Map**
- **Safety in the Science Classroom**
- **Vocabulary Flash Cards**
- **Glossary**
- **Index**

Name _____

Bubble Map

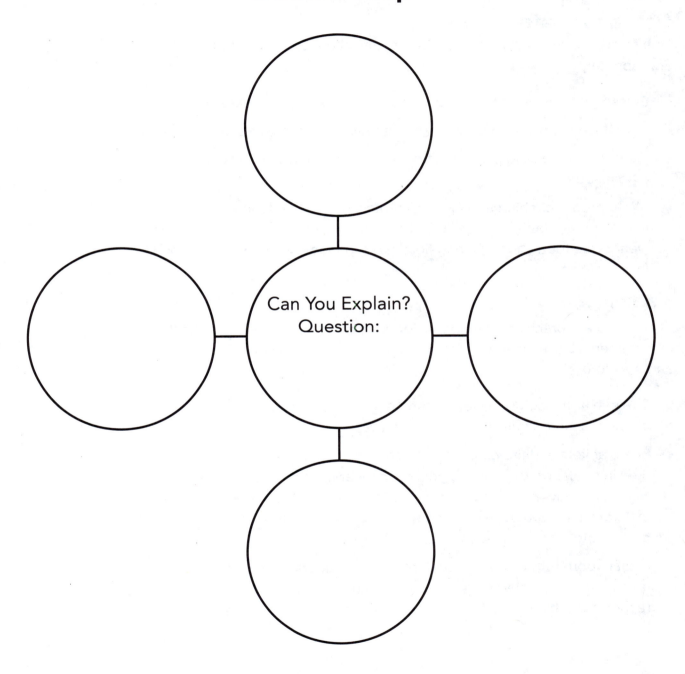

Can You Explain?
Question:

Safety in the Science Classroom

Following common safety practices is the first rule of any laboratory or field scientific investigation.

Dress for Safety

One of the most important steps in a safe investigation is dressing appropriately.

- Splash goggles need to be kept on during the entire investigation.

- Use gloves to protect your hands when handling chemicals or organisms.

- Tie back long hair to prevent it from coming in contact with chemicals or a heat source.

- Wear proper clothing and clothing protection. Roll up long sleeves, and if they are available, wear a lab coat or apron over your clothes. Always wear close toed shoes. During field investigations, wear long pants and long sleeves.

Be Prepared for Accidents

Even if you are practicing safe behavior during an investigation, accidents can happen. Learn the emergency equipment location in your classroom and how to use it.

- The eye and face wash station can help if a harmful substance or foreign object gets into your eyes or onto your face.

- Fire blankets and fire extinguishers can be used to smother and put out fires in the laboratory. Talk to your teacher about fire safety in the lab. He or she may not want you to directly handle the fire blanket and fire extinguisher. However, you should still know where these items are in case the teacher asks you to retrieve them.

- Most importantly, when an accident occurs, immediately alert your teacher and classmates. Do not try to keep the accident a secret or respond to it by yourself. Your teacher and classmates can help you.

Practice Safe Behavior

There are many ways to stay safe during a scientific investigation. You should always use safe and appropriate behavior before, during, and after your investigation.

Safety Goggles

- Read the all of the steps of the procedure before beginning your investigation. Make sure you understand all the steps. Ask your teacher for help if you do not understand any part of the procedure.

- Gather all your materials and keep your workstation neat and organized. Label any chemicals you are using.

- During the investigation, be sure to follow the steps of the procedure exactly. Use only directions and materials that have been approved by your teacher.

- Eating and drinking are not allowed during an investigation. If asked to observe the odor of a substance, do so using the correct procedure known as wafting, in which you cup your hand over the container holding the substance and gently wave enough air toward your face to make sense of the smell.

- When performing investigations, stay focused on the steps of the procedure and your behavior during the investigation. During investigations, there are many materials and equipment that can cause injuries.

- Treat animals and plants with respect during an investigation.

- After the investigation is over, appropriately dispose of any chemicals or other materials that you have used. Ask your teacher if you are unsure of how to dispose of anything.

- Make sure that you have returned any extra materials and pieces of equipment to the correct storage space.

- Leave your workstation clean and neat. Wash your hands thoroughly.

air

Image: Discovery Communications, Inc.

the part of the atmosphere closest to Earth; the part of the atmosphere that organisms on Earth use for respiration

conduction

Image: Discovery Communications, Inc.

when energy moves directly from one object to another

conserve

Image: Emilian Robert Vicol/Pixabay

to protect something, or prevent the wasteful overuse of a resource

Earth

Image: NASA

the third planet from the Sun; the planet on which we live

ecosystem

Image: Paul Fuqua

all the living and nonliving things in an area that interact with each other

energy source

Image: Paul Fuqua

the origin of a form of energy

energy transfer

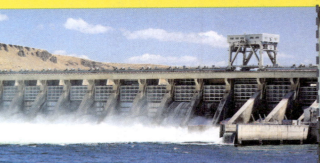

Image: Russ McElroy/Pixabay

the transfer of energy from one object to another, such as heat energy

environment

Image: Odua Images / Shutterstock.com

all the living and nonliving things that surround an organism

fossil fuel

Image: Paul Fuqua

fuels formed from the decay of buried organisms that are pressed tightly together at high temperatures for millions of years

fuel

Image: Paul Fuqua

any material that can be used for energy

generate

Image: Paul Fuqua

to produce by turning a form of energy into electricity

geothermal

Image: WikiImages/Pixabay

heat found deep within the Earth

nonrenewable

Image: Paul Fuqua

once it is used, it cannot be made or reused again

nonrenewable resource

Image: Paul Fuqua

a natural resource of which a finite amount exists, or one which cannot be replaced with currently available technologies

pollution

Image: Paul Fuqua

the introduction of any harmful substance or form of energy into the environment at a faster rate than it can be cleansed.

radiant energy

Image: NASA

energy that does not need matter to travel; light

radiation

Image: Discovery Communications, Inc.

electromagnetic energy

remote (adj)

N-studio / Shutterstock.com

to be operated from a distance

renewable

to reuse or make new again

renewable resource

Image: Paul Fuqua.

a natural resource that can be replaced

resource

Image: Pexels/Pixabay

a naturally occurring material in or on Earth's crust or atmosphere of potential use to humans

sound

Image: Paul Fuqua

a vibration that travels through a material, such as air or water; something that you sense with your hearing

sun

Image: Paul Fuqua

any star around which planets revolve

English ———— **A** ———— **Español**

English	Español
acceleration to increase speed	**aceleración** aumentar la velocidad
adaptation something a plant or animal does to help it survive in its environment (related word: adapt)	**adaptación** proceso mediante el cual las características de una especie cambian a través de varias generaciones como respuesta al medio ambiente (palabra relacionada: adaptar)
air the part of the atmosphere closest to Earth; the part of the atmosphere that organisms on Earth use for respiration	**aire** parte de la atmósfera más cercana a la Tierra; la parte de la atmósfera que los organismos que habitan la Tierra utilizan para respirar
amplitude height or "strength" of a wave	**amplitud** altura o "fuerza" de una onda
analog one continuous signal that does not have any breaks	**analógico** una señal continua que no tiene ninguna interrupción

antenna
a device that receives radio waves and television signals

antena
un dispositivo que recibe ondas de radio y señales de televisión

Arctic
being from an icy climate, such as the north pole

ártico
que tiene relación con el Polo Norte o el área que lo rodea

B

behavior
all of the actions and reactions of an animal or a person (related word: behave)

conducta
todas las acciones y reacciones de un animal (palabra relacionada: comportarse)

brain
the main control center in an animal body; part of the central nervous system

cerebro
principal centro de control en el cuerpo de un animal; parte del sistema nervioso central

C

camouflage
the coloring or patterns on an animal's body that allow it to blend in with its environment

camuflaje
color o apariencia del cuerpo de un animal que le permite mezclarse con su medioambiente

canyon

a deep valley carved by flowing water

cañón

valle profundo esculpido por el flujo del agua

---

chemical energy

energy that can be changed into motion and heat

energía química

energía que está almacenada en las cadenas entre átomos

---

chemical weathering

changes to rocks and minerals on Earth's surface that are caused by chemical reactions

meteorización química

cambios en las rocas y minerales de la superficie de la Tierra causados por reacciones químicas

---

code

a way to communicate by sending messages using dots and dashes

código

una forma de comunicarse enviando mensajes con puntos y rayas

---

collision

the moment where two objects hit or make contact in a forceful way

colisión

el momento en el que dos objetos chocan o hacen contacto de forma contundente

conduction

when energy moves directly from one object to another

conducción

cuando la energía se mueve directamente de un objeto a otro

conservation of energy

energy can not be created or destroyed, it can only be changed from one form to another, such as when electrical energy is changed into heat energy

conservación de la energía

la energía no se puede crear o destruir; solo se puede cambiar de una forma a otra, como cuando la energía eléctrica cambia a energía térmica

conserve

to protect something, or prevent the wasteful overuse of a resource

conservar

proteger algo o evitar el uso excesivo e ineficiente de un recurso

convert (v)

to change forms

convertir (v)

cambiar de forma

Discovery
EDUCATION

delta

a fan-shaped mass of mud and other sediment that forms where a river enters a large body of water

delta

masa de barro y otros sedimentos parecida a un abanico, que se forma donde un río ingresa a un gran cuerpo de agua

deposition

laying sediment back down after erosion moves it around

sedimentación

volver a depositar sedimentos una vez que la erosión los arrastra

digestive system

the body system that breaks down food into tiny pieces so that the body's cells can use it for energy

sistema digestivo

sistema del cuerpo que divide los alimentos en pequeños trozos para que las células del cuerpo puedan usarlos para obtener energía.

digital

a signal that is not continuous and is made up of tiny separate pieces

digital

una señal que no es continua y está compuesta por diminutas partes separadas

energy transfer

the transfer of energy from one organism to another through a food chain or web; or the transfer of energy from one object to another, such as heat energy

transferencia de energía

transmisión de energía desde un organismo a otro a través de una cadena o red de alimentos; o transferencia de energía desde un objeto a otro, como por ejemplo la energía del calor

engineer

Engineers have special skills. They design things that help solve problems.

ingeniero

Los ingenieros poseen habilidades especiales. Diseñan cosas que ayudan a resolver problemas.

environment

all the living and nonliving things that surround an organism

medio ambiente

todos los seres vivos y objetos sin vida que rodean a un organismo

erosion

the removal of weathered rock material. After rocks have been broken down, the small particles are transported to other locations by wind, water, ice, and gravity

erosión

proceso por el cual el viento, el agua, el hielo y otras cosas mueven trozos de roca y suelo sobre la superficie de la Tierra

Discovery EDUCATION

erupt

the action of lava coming out of a hole or crack in Earth's surface; the sudden release of hot gasses or lava built up inside a volcano (related word: eruption)

erupción

acción de la lava que sale de un agujero o cráter de la superficie de la Tierra; repentina liberación de gases o lava calientes formados en el interior de un volcán (palabra relacionada: erupción)

extinct

describes a species of animals that once lived on Earth but which no longer exists (related word: extinction)

extinto

palabra que hace referencia a una especie de animales que una vez habitó la Tierra, pero que ya no existe (palabra relacionada: extinción)

— F —

fault

a fracture, or a break, in the Earth's crust (related word: faulting)

falla

grieta en el cuerpo de una roca que hace que la roca se desplace (palabra relacionada: fallas)

feature

things that describe what
something looks like

rasgo

cosas que describen cómo se ve
algo

force

a pull or push that is applied to
an object

fuerza

acción de atraer o empujar que
se aplica a un objeto

forecast

(v) to analyze weather data
and make an educated guess
about weather in the future;
(n) a prediction about what the
weather will be like in the future
based on weather data

pronosticar / pronóstico

(v) analizar los datos del tiempo
y hacer una conjetura informada
sobre el tiempo en el futuro; (s)
predicción sobre cómo será el
tiempo en el futuro en base a
datos

fossil fuels

fuels that come from very old life
forms that decomposed over a
long period of time, like coal, oil,
and natural gas

combustibles fósiles

varios tipos de combustibles
formados de manera natural a
partir de los restos de plantas
y animales que murieron hace
miles o millones de años

friction

a force that slows down or stops motion

fricción

fuerza que se opone al movimiento de un cuerpo sobre una superficie o a través de un gas o un líquido

---

fuel

any material that can be used for energy

combustible

todo material que puede usarse para producir energía

 G ───────────

generate

to produce by turning a form of energy into electricity

generar

producir convirtiendo una forma de energía en electricidad

---

geothermal

heat found deep within Earth

geotérmica

calor que se encuentra en la profundidad de la Tierra

---

glacier

a large sheet of ice or snow that moves slowly over Earth's surface

glaciar

ran sábana de hielo o nieve que se mueve lentamente sobre la superficie de la Tierra

gravitational potential energy
energy stored in an object based on its height and mass

energía potencial gravitacional
energía almacenada debida a la ubicación en un campo gravitacional

gravity
the force that pulls an object toward the center of Earth (related word: gravitational)

gravedad
fuerza que existe entre dos objetos cualquiera que tienen masa (palabra relacionada: gravitacional)

— **H** —

heart
the muscular organ of an animal that pumps blood throughout the body

corazón
órgano muscular de un animal que bombea sangre a través del cuerpo

heat
the transfer of thermal energy

calor
transferencia de energía térmica

hibernate
to reduce body movement during the winter in an effort to conserve energy (related word: hibernation)

hibernar
reducir el movimiento del cuerpo durante el invierno con la finalidad de conservar la energía (palabra relacionada: hibernación)

— I —

information
facts or data about something; the arrangement or sequence of facts or data

información
hechos o datos sobre algo; la organización o secuencia de hechos o datos

— K —

kinetic energy
the energy an object has because of its motion

energía cinética
energía que posee un objeto a causa de su movimiento

landform

a large natural structure on Earth's surface, such as a mountain, a plain, or a valley

accidente geográfico

estructura natural grande que se encuentra en la superficie de la Tierra, como una montaña, una llanura o un valle

lava

molten rock that comes through holes or cracks in Earth's crust that may be a mixture of liquid and gas but will turn into solid rock once cooled

lava

roca fundida que sale por orificios o grietas en la corteza terrestre, y que puede ser una mezcla de líquido y gas pero se convierte en roca sólida al enfriarse

light

a form of energy that moves in waves and particles and can be seen

luz

ondas de energía electromagnéticas; energía electromagnética que la gente puede ver

magma
melted rock located beneath Earth's surface

magma
roca fundida que se encuentra debajo de la superficie de la Tierra

magnetic field
a region in space near a magnet or electric current in which magnetic forces can be detected

campo magnético
región en el espacio cerca de un imán o de una corriente eléctrica, donde pueden detectarse fuerzas magnéticas

map
a flat model of an area

mapa
modelo plano de un área

mass
the amount of matter in an object

masa
cantidad de materia en un objeto

matter
material that has mass and takes up some amount of space

materia
material que tiene masa y ocupa cierta cantidad de espacio

meander
winding or indirect movement or course

meandro
movimiento o curso serpenteante o indirecto

migration

the movement of a group of organisms from one place to another, usually due to a change in seasons

migración

movimiento de un grupo de organismos de un lugar a otro, generalmente debido a un cambio de estaciones

model

a drawing, object, or idea that represents a real event, object, or process

modelo

dibujo, objeto o idea que representa un evento, objeto, o proceso real

motion

when something moves from one place to another (related words: move, movement)

movimiento

cambio en la posición de un objeto en comparación con otro objeto (palabra relacionada: mover, desplazamiento)

mountain

an area of land that forms a peak at a high elevation (related term: mountain range)

montaña

área de tierra que forma un pico a una elevación alta (palabra relacionada: cadena montañosa)

nerve

a cell of the nervous system that carries signals to the body from the brain, and from the body to the brain and/or spinal cord

nervio

célula del sistema nervioso que lleva señales al cuerpo desde el cerebro, y desde el cuerpo al cerebro y/o médula espinal

nonrenewable

once it is used, it cannot be made or reused again

no renovable

no renovable

nonrenewable resource

a natural resource of which a finite amount exists, or one that cannot be replaced with currently available technologies

recurso no renovable

recurso natural del cual existe una cantidad finita, o uno que no puede remplazarse con las tecnologías actualmente disponibles

nuclear energy

the energy released when the nucleus of an atom is split apart or combined with another nucleus

energía nuclear

energía liberada cuando el núcleo de un átomo se divide o combina con otro átomo

O

ocean
a large body of salt water that covers most of Earth

océano
gran cuerpo de agua salada que cubre la mayor parte de la Tierra

opaque
describes an object that light cannot travel through

opaco
describe un objeto que la luz no puede atravesar

organ
a group of tissues that performs a complex function in a body

órgano
conjunto de tejidos que realizan una función compleja en el cuerpo

organism
any individual living thing

organismo
todo ser vivo individual

P

photosynthesis
the process in which plants and some other organisms use the energy in sunlight to make food

fotosíntesis
proceso en el cual las plantas y algunos otros organismos usan la energía del Sol para producir alimentos

pollute

to put harmful materials into the air, water, or soil (related words: pollution, pollutant)

contaminar

poner materiales perjudiciales en el aire, agua o suelo (palabras relacionadas: contaminación, contaminante)

pollution

when harmful materials have been put into the air, water, or soil (related word: pollute)

contaminación

cuando se introducen materiales perjudiciales en el aire, el agua o el suelo (palabra relacionada: contaminar)

potential energy

the amount of energy that is stored in an object; energy that an object has because of its position relative to other objects

energía potencial

cantidad de energía almacenada en un objeto; energía que tiene un objeto debido a su posición relativa con otros objetos

predator

an animal that hunts and eats another animal

depredador

animal que caza y come a otro animal

predict
to guess what will happen in the future (related word: prediction)

predecir
adivinar qué sucederá en el futuro (palabra relacionada: predicción)

prey
an animal that is hunted and eaten by another animal

presa
animal que es cazado y comido por otro

pupil
the black circle at the center of an iris that controls how much light enters the eye

pupila
círculo negro en el centro del iris que controla cuánta luz entra al ojo

 R

radiant energy
energy that does not need matter to travel; light

energía radiante
energía que no necesita de la materia para viajar; luz

radiation
electromagnetic energy (related word: radiate)

radiación
energía electromagnética (palabra relacionada: irradiar)

receptor

nerves located in different parts of the body that are especially adapted to receive information from the environment

receptor

nervios ubicados en diferentes partes del cuerpo que están especialmente adaptados para recibir información del medio ambiente

reflect

light bouncing off a surface (related word: reflection)

reflejar

golpear sobre una superficie y rebotar en la dirección opuesta (palabra relacionada: reflexión)

reflex

an automatic response

reflejo

respuesta automática

refract

to bend light as it passes through a material (related word: refraction)

refractar

inclinación de la luz cuando pasa a través de un material (palabra relacionada: refracción)

remote (adj)

to be operated from a distance

remoto (adj)

que se opera a distancia

renewable

to reuse or make new again

renovable

reutilizar o volver a hacer de nuevo

---

renewable resource

a natural resource that can be replaced

recurso renovable

recurso natural que puede reemplazarse

---

reproduce

to make more of a species; to have offspring (related word: reproduction)

reproducir

hacer más de una especie; tener descendencia (palabra relacionada: reproducción)

---

resistance

when materials do not let energy transfer through them

resistencia

cuando los materiales no permiten la transferencia de energía a través de ellos

---

resource

a naturally occurring material in or on Earth's crust or atmosphere of potential use to humans

recurso

material que se origina de forma natural en o sobre la corteza o la atmósfera de la Tierra, que es de uso potencial para los seres humanos

rotate

turning around on an axis;
spinning (related word: rotation)

rotar

girar sobre un eje; dar vueltas
(palabra relacionada: rotación)

--------- S ---------

satellite

a natural or artificial object that
revolves around another object
in space

satélite

objeto natural o artificial que gira
alrededor de otro objeto en el
espacio

sediment

solid material, moved by wind
and water, that settles on the
surface of land or the bottom of a
body of water

sedimento

material sólido que el viento o el
agua mueve y que se asienta en
la superficie de la tierra o en el
fondo de un cuerpo de agua

seismic

having to do with earthquakes or
earth vibrations

sísmico

relativo a los terremotos o a las
vibraciones de la Tierra

seismic wave

waves of energy that travel through the Earth

onda sísmica

ondas de energía que viajan a través del interior de la Tierra debido a un terremoto, otras fuerzas tectónicas o una explosión

senses

taste, touch, sight, smell, and hearing (related word: sensory)

sentidos

gusto, tacto, visión, olfato y audición (palabra relacionada: sensorial)

skin

an organ that covers and protects the bodies of many animals

piel

órgano que cubre y protege los cuerpos de muchos animales

soil

the outer layer of Earth's crust in which plants can grow; made of bits of dead plant and animal material as well as bits of rocks and minerals

suelo

capa externa de la corteza de la Tierra en donde crecen las plantas; formada por pedazos de plantas y animales muertos, así como por pedazos de rocas y de minerales

Discovery
EDUCATION

solar energy

energy that comes from the sun

energía solar

energia que proviene del Sol

sound

anything you can hear that travels by making vibrations in air, water, and solids

sonido

vibración que viaja a través de un material, como el aire o el agua; lo que se percibe a través de la audición

sound wave

a sound vibration as it is passing through a material: Most sound waves spread out in every direction from their source.

onda sonora

vibración de sonido que se produce cuando se atraviesa un material: la mayoría se dispersa desde la fuente en todas direcciones.

speed

the measurement of how fast an object is moving

velocidad

distancia recorrida por unidad de tiempo

stimulus

things in the environment that cause us to react or have a physical response

estímulo

algo en el medio ambiente que nos hace reaccionar o tener una respuesta física

stomach

a muscular organ in the body where chemical and mechanical digestion take place

estómago

órgano muscular del cuerpo donde tiene lugar la digestión química y mecánica

sun

any star around which planets revolve

sol

toda estrella alrededor de la cual giran los planetas

survive

to continue living or existing: an organism survives until it dies; a species survives until it becomes extinct (related word: survival)

sobrevivir

continuar viviendo o existiendo: un organismo sobrevive hasta que muere; una especie sobrevive hasta que se extingue (palabra relacionada: supervivencia)

system

a group of related objects that work together to perform a function

sistema

grupo de objetos relacionados que funcionan juntos para realizar una función

tectonic plate
one of several huge pieces of Earth's crust

placa tectónica
una de las muchas piezas enormes de la corteza terrestre

thermal energy
energy in the form of heat

energía térmica
energía en forma de calor

tongue
an organ in the mouth that helps in eating and speaking

lengua
órgano de la boca que ayuda a comer y hablar

topographic map
a map that shows the size and location of an area's features such as vegetation, roads, and buildings

mapa topográfico
mapa que muestra el relieve y otras características de un área

trait
a characteristic or property of an organism

rasgo
característica o propiedad de un organismo

transparent

describes materials through which light can travel; materials that can be seen through

transparente

describe materiales a través de los cuales puede viajar la luz; materiales a través de los cuales se puede ver

---

tsunami

a giant ocean wave (related word: tidal wave)

tsunami

hola oceánica gigante (palabra relacionada: maremoto)

valley

a low area of land between two higher areas, often formed by water

valle

área baja de tierra entre dos áreas más altas, generalmente formada por el agua

---

volcano

an opening in Earth's surface through which magma and gases or only gases erupt (related word: volcanic)

volcán

abertura en la superficie de la Tierra a través de la cual el magma y los gases o sólo los gases hacen erupción (palabra relacionada: volcánico)

water

a compound made of hydrogen and oxygen; can be in either a liquid, ice, or vapor form and has no taste or smell

agua

compuesto formado por hidrógeno y oxígeno

wave

a disturbance caused by a vibration; waves travel away from the source that makes them

onda

perturbación causada por una vibración que se aleja de la fuente que la forma

wavelength

the distance between one peak and the next on a wave

longitud de onda

distancia entre un pico y otro en una onda

weathering

the physical or chemical breakdown of rocks and minerals into smaller pieces or aqueous solutions on Earth's surface

meteorización

desintegración física o química de rocas y minerales en pedazos más pequeño o en soluciones acuosas en la superficie de la Tierra

work

a force applied to an object over a distance

trabajo

fuerza aplicada a un objeto a lo largo de una distancia

Index